Computational Methods in Applied Sciences

Series Editor: Eugenio Oñate

Jan Holnicki-Szulc · Carlos A. Mota Soares (Eds.)

Advances in Smart Technologies in Structural Engineering

Springer

Series Editor:
Prof. Dr. Eugenio Oñate
Universidad Politécnica de Cataluna
Intern. Center for Numerical
Methods in Engineering (CIMNE)
Edifico C-1, Campus Norte U.P.C
Gran Capitan, s/n
08034 Barcelona
Spain

Editors:
Prof. Dr. Jan Holnicki-Szulc
Inst. Fundamental Technological Research
Polish Academy of Sciences
ul. Swietokrzyska 21,
00–049 Warszawa
Poland

Prof. Dr. Carlos Mota Soares
Universidade Tecnica Lisboa
Inst. Superior Tecnico
Centro Mecanica e Materiais
Avenida Rivusci Pais
1049–001 Lisboa
Portugal

ISBN 3-540-22331-2 Springer-Verlag Berlin Heidelberg New York

Cataloging-in-Publication Data applied for

Bibliographic information published by Die Deutsche Bibliothek
Die Deutsche Bibliothek lists this publication in the Deutsche Nationalbibliografie;
detailed bibliographic data is available in the Internet at <http://dnb.ddb.de>.

Printed in Germany
springeronline.com

Coverdesign: Design and Production, Heidelberg
62/3020 UW Printed on acid-free paper – 5 4 3 2 1 0 –

AMAS/ECOMAS/STC Workshop on
Smart Materials and Structures

ECCOMAS

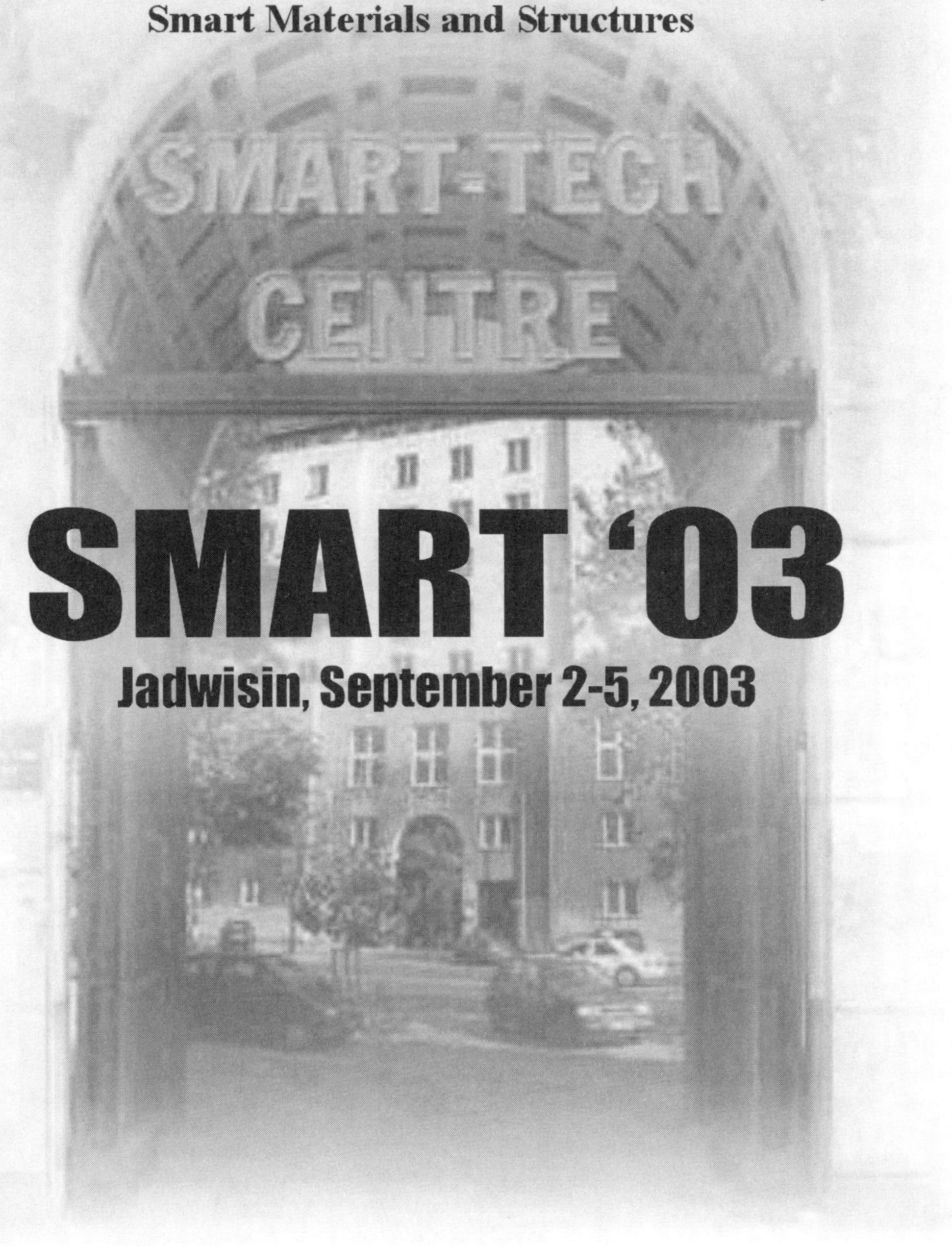

SMART '03

Jadwisin, September 2-5, 2003

Foreword

This book collects invited lectures presented and discussed on the AMAS & ECCOMAS Workshop/Thematic Conference SMART'03. The SMART'03 Conference on Smart Materials and Structures was held in a 19th century palace in Jadwisin near Warsaw, 2-5 September 2003, Poland .It was organized by the Advanced Materials and Structures (AMAS) Centre of Excellence at the Institute of Fundamental Technological Research (IFTR) in Warsaw, ECCOMAS – European Community on Computational Methods in Applied Sciences and SMART-TECH Centre at IFTR. The idea of the workshop was to bring together and consolidate the community of Smart Materials and Structures in Europe. The workshop was attended by 66 participants from 11 European countries (Austria, Belgium, Finland, France, Germany, Italy, Poland, Portugal, Spain, U.K., Ukraine), 1 participant from Israel and 1 participant from the USA.

The workshop program was grouped into the following major topics:

- 4 sessions on **Structural Control** (18 presentations),
- 3 sessions on **Vibration Controland Dynamics** (14 presentations),
- 2 sessions on **Damage Identification** (10 presentations),
- 2 sessions on **Smart Materials** (9 presentations).

Each session was composed of an invited lecture and some contributed papers. Every paper scheduled in the program was presented, so altogether 51 presentations were given. No sessions were run in parallel. The workshop was attended not only by researchers but also by people closely related to the industry. There were interesting discussions on scientific merits of the presented papers as well as on future development of the field and its possible industrial applications.

Apart from the sessions there was also a one-afternoon exhibition on smart materials and devices prepared by commercial companies working in the field, i.e. LORD Corp. (USA), CEDRAT (France) and by the Wroclaw University of Technology (Poland). Commercial and academic applications of smart materials (magneto-rheological, piezo-electric, magneto-strictive) and software tools relevant for the field were demonstrated.

Jan Holnicki-Szulc
Carlos Mota-Soares
Co-Chairmen

Table of Contents

Fuzzy Chip Controllers and Wireless Links in Smart Structures

Fabio Casciati and Roberto Rossi

Dept. of Structural Mechanics,
via Ferrata, 1, 27100 Pavia, Italy

Summary. This paper summarizes some authors' developments on the design and implementation of fuzzy chip controllers and wireless sensor networks. The design of a fuzzy controller is first discussed in detail. Then a section presents the embedded software that has been developed for the microcontroller. A few considerations on how to provide the system with a wireless networking capability are provided and two architectures are discussed. Extensive tests of the fuzzy controller are shown in order to report the advances achieved in this field.

1 Introduction

In the design of an active (hybrid) control system, the aim is the reduction of the structural response in terms of accelerations, velocities and displacements under the limitation of both the control force level (limited by the actuator features and by the required amount of energy) and the number of measured signals.

When dealing with civil engineering structures, one has to meet several constraints in conceiving semi-active or active control solutions:

1. Reliable numerical models of the structural system are generally required. They will be used to simulate and predict both the structural response and the control system performance. Only a few classes of controllers do not pose this requirement, and among them the fuzzy controllers.
2. The structure behavior must be available in real time thanks to a continuous monitoring of the structural system.
3. A wide spectrum of control algorithms must be considered. The one to be selected strongly depends on the application.
4. The control system must show robustness in order to adapt to system modifications (loads, cracking, etc.).
5. Devices able to produce high forces with minimal power requirements must be available. Power should be accumulated before operation, without relying on its external supply.

Items from 1 to 3 exploit the expertise gained in many full scale structures which were monitored in order to investigate their *health* and to detect possible damages [2, 4]. But the extension of some current developments, like wireless sensor networks, must be carefully investigated [8, 12].

Item 4 becomes the most critical for control systems in civil engineering, because the scenarios to be controlled can be predicted and estimated, but they will never be completely known *a priori*. Item 5 is quite important: it is far from the purposes of this paper, but a wide literature covers the topic [3].

Fuzzy theory was deeply investigated in view of civil engineering applications [13]. Fuzzy control theory, an alternative to classical control theory, allows the resolution of imprecise or uncertain information [6, 7] in order to overcome the problems already discussed in items 3 and 4.

This paper summarizes some authors' developments on the design and implementation of fuzzy chip controllers and wireless sensor networks. The core of the paper is a board designed and assembled by the authors. It is illustrated in the next section.

2 Board Description

The board supports four analog inputs and one analog output. In addition, a serial port compatible with the EIA-232 standard is available and is used as an interface for command input from a PC or any other compliant device.

The heart of the board is a C8051F007 microcontroller from Cygnal Integrated Products Inc. It is an 8051-based microprocessor with on-chip peripherals (ADC, DAC, UART, etc.) and flash memory. The maximum clock frequency is 25MHz, which is perfectly suited to processing signals whose spectrum spans from 0 to 25 Hz. It has 32Kbytes in-system-programmable flash memory plus on-chip RAM in excess of 2Kbytes.

The 8051 architecture is very widely used in the industry and is very flexible in that it supports any type of peripheral. Many 8051-based microcontrollers exist on the market with different features. Those from Cygnal Integrated Products were among the first to implement a high speed core capable of executing most of the instructions in a single clock cycle, as opposed to the standard 8051 core that executes most of its instructions in no less than 12 clock cycles. Moreover, at the time this board was designed, these microcontrollers were those with the greatest amount of on-chip flash memory.

Microcontrollers based on the 8051 core are standard and can be programmed using standard and well-tested C compilers available from many vendors. This is the main reason they have been chosen for this board. The C8051F007 has a peak 25 million instructions per second, which is enough for implementing a fuzzy controller as well as a linear controller. Although this board has been mainly designed with fuzzy control in mind, it is suitable for hosting any type of controller as long as no more than 25 million instructions per second are required.

Previous work [9, 10] was based on a microprocessor with hardware fuzzy support. However, the non-volatile memory was EPROM, not flash, thus lacking in flexibility. The new version of this chip, although equipped with flash memory, was not adopted for a few reasons. First, the highest ADC resolution is 10 bits. Second, the chip is not 8051-compatible, which means software porting to and from other hardware solutions is not as easy. Third, chips from Analog Devices and Cygnal are promptly delivered and are easier to buy. A drawback of this choice could be the lack of support for fuzzy logic, but it is very easy to implement fuzzy logic in C language, so this is not much of a problem actually.

Figure 1 shows the top-level board diagram of the fuzzy controller board. The microcontroller is connected to the input channels, the output channel, the RS-232 interface and the JTAG interface.

The input channels perform offset voltage cancellation by means of a simple first-order high-pass filter made up of a capacitor and a resistor. The corner frequency of the high-pass network is 0.016Hz. This means the phase distortion introduced at 0.5Hz is 1.8 degrees, while the amplitude distortion is less than 0.005dB. Same figures at 1Hz are 0.9 degrees and 0.001dB, respectively. This means the amount of distortion introduced in the signal band (i.e., 0.5Hz $< f <$ 25Hz) is negligible.

The output channel has the same offset cancellation filter and, in addition, has the capability of setting the dc output value either to 0V or to 5V. This

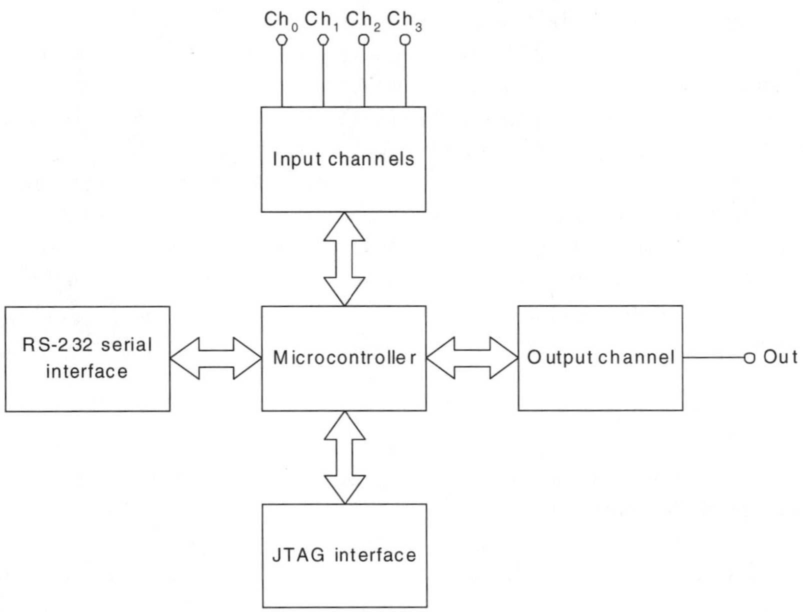

Fig. 1. Top-level board diagram

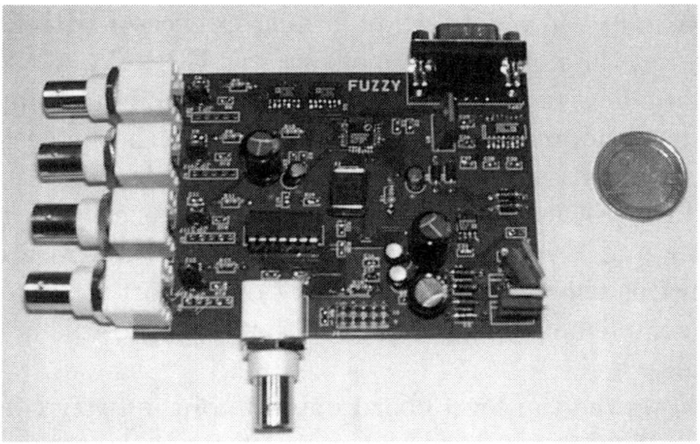

Fig. 2. Photograph of the board

way, two output ranges are possible: 0V to 10V and −5V to +5V. Output range selection is carried out through software.

The RS-232 interface allows the board to communicate with a computer or with another board of the same kind or even with a different peripheral such as, for example, a radio modem or another kind of wireless transceiver. This issue will be covered in a later section.

The board needs a dual ±15V power supply or higher. Voltage regulation is carried out on board. In case, the output channel is not used, the board can function out of a simple 9V battery. In this case, only the input channels and the serial port will be operational. With suitable software, in this mode the board could, for example, still work as a controller and drive an actuator through a wireless link (e.g., a radio modem connected to the serial port).

The microcontroller works out of a 3.3V power supply. A 25MHz quartz crystal provides the chip with a stable and accurate clock source.

Finally, the JTAG interface is used for downloading the software into the flash memory of the chip and for debugging purposes. A software tool from Cygnal, along with a serial-to-JTAG converter (a small and inexpensive black box), provides programming and in-system, real-time debugging facilities.

Fig. 2 is a photograph of the fuzzy controller board.

3 Board Software

The software has been designed with ease of use as main concern. It has been implemented as a terminal shell. The board can be connected to any device hosting a standard serial port. In the EIA-232 standard jargon, the board acts as a DTE (data terminal equipment) rather than a DCE (data communication equipment) and, therefore, acts like a personal computer rather than a modem. Hence, it can be connected to a personal computer using a null-modem serial cable.

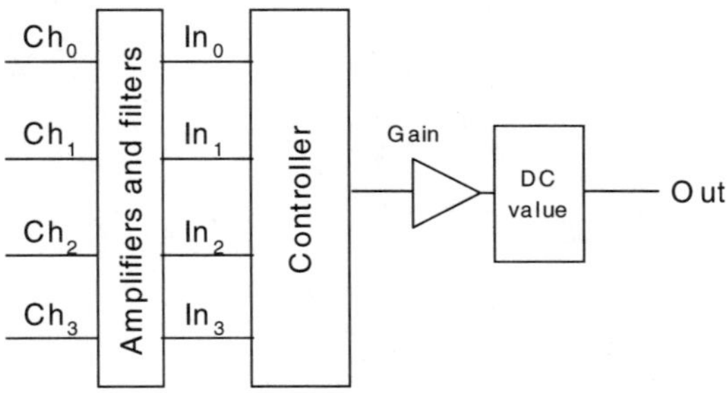

Fig. 3. Equivalent architecture of the embedded software

On the computer, a software like HyperTerminal (or any other equivalent terminal emulation software) can be used to access the embedded shell interface exposed by the board. The connection parameters are as follows: baud-rate = 38400, parity = none, stop bits = 1, flow control = none.

The on-chip software is equivalent to the block diagram in Figure 3.

The four analog inputs, which have a 12-bit resolution, are first passed through programmable amplifiers and programmable filters. The gains of the amplifiers can be assigned through the `in` command of the terminal shell, while the filters can be programmed via the shell command `filter`: Four filters are available:

- no filter,
- offset cancellation,
- approximate integration,
- approximate integration followed by offset cancellation.

The values obtained after filtering are the actual inputs to the controller. The type of controller is user-selectable (command `control`). The available controllers are:

- no control,
- sinusoid,
- linear controller,
- fuzzy controller.

The output of the controller is then amplified by a programmable factor (command `out`) and sent out of the chip through the 12-bit digital-to-analog converter. The desired mean value of the output channel is also user-selectable (command `out`).

The shell commands currently available are:

- acquire
- control
- desc
- filter

- fuzzy
- help
- in
- load
- monitor
- out
- save
- version
- sinusoid
- linear

The commands are described in detail in the following section.

3.1 General Commands

help

With no parameters lists all the available shell commands.
If followed by a command name among those listed, it provides a brief help on how to use the specified command.

Examples:

```
help
help acquire
```

version

Gives the current on-chip software version.

acquire

Enters acquire mode. Pressing a key when in acquire mode will cause the chip to leave acquire mode back into normal mode. In normal mode, the shell accepts commands. In acquire mode the shell will output a line at every sampling interval. The sampling frequency is 50Hz and therefore the shell will print 50 lines per second. Each line is composed of 11 columns:

1. Timestamp (the number of sampling intervals elapsed since power on); Timestamp/50 gives the number of seconds elapsed since the board was powered on
2. Ch0
3. Ch1
4. Ch2
5. Ch3
6. In0
7. In1
8. In2
9. In3
10. Out
11. ChOut

If the board is turned off while in acquire mode, it will reenter acquire mode when powered up again. If the board is turned off while in normal mode, it will start up in normal mode. In other words, acquire/normal mode persists even through power on/off cycles.

monitor

Displays the same data of acquire mode but in a more readable form. This command does not persist through power on/off cycles.

3.2 Load & Save

desc

With no parameters: prints the current short description.
Followed by a string: associates a short description (20 chars) to the current system configuration. This description is loaded and saved by load and save. It is useful for retrieving a configuration saved before.

Examples:

```
desc
desc Awesome Controller
```

load

With no parameters: lists all the saved configurations.
With a slot number (0-9): loads the configuration from the slot specified. The configurations are saved in a reserved portion of the chip flash memory.
With a slot number and an asterisk: loads the specified slot and sets it as the default slot. The default slot is automatically loaded when the board is powered on.

Examples:

```
load
load 5
load 4 *
```

save

With a slot number (0-9): saves the current configuration into the specified slot in the reserved portion of the flash memory. A short description should be set (command desc) before saving into a slot. This will allow configurations to be easily retrieved at a later time.

Example:

```
save 0
```

3.3 Control Type and Weights

in

Defines the elements of the input matrix A.
Syntax:

```
in <id> w0 w1 w2 w3
```

Defines the (unfiltered) input labeled by <id> as w0*Ch0 + w1*Ch1 + w2*Ch2 + w3*Ch3.
The four weights w0, w1, w2 and w3 are floating-point numbers.

Examples:

```
in 0 3 0 -3 0
```

Defines $\mathrm{In}_{0,\mathrm{unfiltered}}$ = 3*Ch0 - 3*Ch2.

```
in 1 1.1 -3.6 4.3 0
```

Defines $\mathrm{In}_{1,\mathrm{unfiltered}}$ = 1.1*Ch0 - 3.6*Ch1 + 4.3*Ch2 + 0*Ch3

filter

Selects the four filters to be applied to the four controller inputs.
Syntax:

```
filter f0 f1 f2 f3
```

Applies filter f0 to $\mathrm{In}_{0,\mathrm{unfiltered}}$, filter f1 to $\mathrm{In}_{1,\mathrm{unfiltered}}$, etc...
The values f0, f1, f2 and f3 are integers taken from the following list:

1. no filter
2. offset cancellation
3. approximate integration
4. approximate integration followed by offset cancellation

Examples:

```
filter 0 0 0 0
```

Turns off filtering for all the four inputs.

```
filter 0 2 2 1
```

Defines the four controller inputs as follows:

$$\mathrm{In}_0 = \mathrm{In}_{0,\mathrm{unfiltered}}$$
$$\mathrm{In}_1 = \mathrm{ApproxIntegrator}(\mathrm{In}_{1,\mathrm{unfiltered}})$$
$$\mathrm{In}_2 = \mathrm{ApproxIntegrator}(\mathrm{In}_{2,\mathrm{unfiltered}})$$
$$\mathrm{In}_3 = \mathrm{OffsetCanc}(\mathrm{In}_{3,\mathrm{unfiltered}})$$

control

Selects the type of controller to be used. Changing the type of controller will reset all the parameters entered via the fuzzy, linear and sinusoid commands, as the parameters for different controllers are all saved in the same memory area. Therefore, this command should be issued first and should not be entered again after setting the controller parameters or the parameters will likely have to be reentered from scratch.

Syntax:

`control <type>`

The `<type>` parameter is an integer taken from the following list:

0. no control (the output of the controller is always zero, no configuration parameters are available in this case)
1. fuzzy control (configure the controller via the fuzzy command)
2. sinusoid controller (the output is a simple sinusoid regardless of the values of the inputs, configure via the sinusoid command)
3. linear controller (configure via the linear command)

Examples:

```
control 0
control 2
```

out

Sets the output gain and the output mean value.

Syntax:

`out <gain> <mean>`

`<gain>` is a floating-point value, `<mean>` can be either 0 or 5 (volt).

Examples:

`out -1 0`

Gain –1 and mean value = 0V

`out 1.2 5`

Gain = 1.2 and mean value = 5V

3.4 Controller Parameters

fuzzy

Sets the controller parameters in case the fuzzy controller is selected. This command allows fuzzy sets and rules to be defined.

When invoked with no parameters, this command simply lists all the fuzzy sets and rules currently defined. Otherwise, the general syntax of this command is:

`fuzzy <name of thing to configure> <parameters>...`

Syntax 1 (general configuration):

`fuzzy config <n0> <n1> <n2> <n3> <nOut> <nRules>`

Configures the fuzzy controller to use `n0` fuzzy sets for input `In0`, `n1` for `In1`, `n2` for `In2`, `n3` for `In3`, `nOut` crisp fuzzy sets for the output variable `Out`, and `nRules` fuzzy rules.

This command actually allocates memory space for sets and rules and, hence, must be invoked before actually defining sets and rules.

Example:

`fuzzy config 3 3 1 1 5 9`

Instructs the fuzzy controller to allocate space for:

- 3 fuzzy sets for In0
- 3 fuzzy sets for In1
- 1 fuzzy set for In2
- 1 fuzzy set for In3
- 5 fuzzy sets for Out
- 9 fuzzy rules

Syntax 2 (input fuzzy sets):

```
fuzzy input <var> <setID> <setName> <A> <B> <C>
```

Defines setName as the triplet A, B, C. This setName is associated to the input variable var and is given the specified setID.
<var> can be 0, 1, 2 or 3 and means In0, In1, In2 or In3, respectively.
<setID> is the zero-based index of the set that is being defined. For example, if the controller has been configured to use 3 fuzzy sets for In0, valid setID's are 0, 1 and 2.
<setName> is a convenience name that is associated to the set and is only used when displaying sets and rules in a human-readable form (the fuzzy command with no parameters).
The triplet A, B, C actually defines the fuzzy set. Input fuzzy sets can be either triangular or trapezoidal. The trapezoidal shape is just a generalization of the triangular shape when one of the vertices is located at $\pm\infty$.
A, B and C can be integers between -32768 and 32767; they can also be assigned the two special values +inf and $-$inf.

Examples:

```
fuzzy input 1 2 PO 0 1000 +inf
```

Defines a fuzzy set for variable In1. The fuzzy set is that with index 2 and is assigned the name PO. Its geometry is trapezoidal, with A=0, B=1000 and C=$+\infty$.

```
fuzzy input 1 1 ZE -1000 0 1000
```

Defines a fuzzy set for variable In1. The fuzzy set is that with index 1 and is assigned the name ZE. Its shape is triangular, with A=-1000, B=0 and C=+1000.

Syntax 3 (output fuzzy sets):

```
fuzzy output <setID> <setName> <A>
```

Defines a crisp fuzzy set called setName, with index setID and located at position A.
<setID> is the zero-based index of the set that is being defined. For example, if the controller has been configured to use 7 fuzzy sets for the output variable Out then valid setID's are 0, 1, 2, 3, 4, 5 and 6.
<setName> is a convenience name that is associated to the set and is only used when displaying sets and rules in a human-readable form (the fuzzy command with no parameters).
The value A actually defines the fuzzy set. Output fuzzy sets are crisp and, therefore, are defined by a single value.

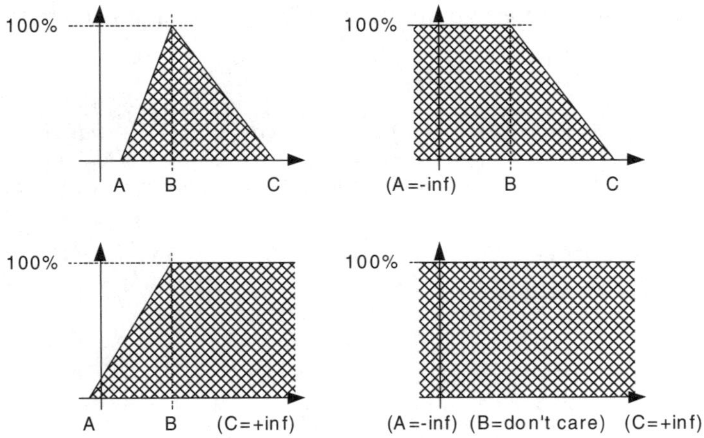

Fig. 4. Meaning of the triplet A, B, C

Example:

`fuzzy output 2 NE -500`

Defines an output fuzzy set called NE with a value of −500. This set is that with index 2 and, therefore, is the third output fuzzy set.

Syntax 4 (fuzzy rules):

`fuzzy rule <ruleID> <s0> <s1> <s2> <s3> <sOut>`

Defines the rule whose index is `ruleID`. Valid values for `ruleID` are 0 to the number of allocated rules (command fuzzy config) minus one. The remaining five parameters are indexes of fuzzy sets.

`<s0>` is the zero-based index of a fuzzy set of the `In0` variable.
`<s1>` is the zero-based index of a fuzzy set of the `In1` variable.
`<s2>` is the zero-based index of a fuzzy set of the `In2` variable.
`<s3>` is the zero-based index of a fuzzy set of the `In3` variable.
`<sOut>` is the zero-based index of a fuzzy set of the `Out` variable.

The meaning of the rule is:

"If `In0` ∈ `s0` and `In1` ∈ `s1` and `In2`∈ `s2` and `In3` ∈ `s3` then `Out` = `sOut`"

Example:

`fuzzy rule 21 3 4 2 1 6`

Defines rule 21 as:

If `In0` ∈ "set 3 of `In0`" and `In1` ∈ "set 4 of `In1`" and `In2` ∈ "set 2 of `In2`" and `In3` ∈ "set 1 of `In3`" then the crisp value to be assigned to `Out` is that of "set 6 of `Out`"

linear

Sets the controller parameters in case the linear controller is selected. This command allows four linear transfer functions to be defined. In the z-domain, the relationship implemented between inputs and output of the controller is:

$$\mathrm{Out}(z) = \frac{n_0(z)}{d(z)} \cdot \mathrm{In}_0(z) + \frac{n_1(z)}{d(z)} \cdot \mathrm{In}_1(z) + \frac{n_2(z)}{d(z)} \cdot \mathrm{In}_2(z) + \frac{n_3(z)}{d(z)} \cdot \mathrm{In}_3(z) \quad (1)$$

The five functions n_0, n_1, n_2, n_3 and d are polynomials in the form:

$$c_0 + c_1 z^{-1} + c_2 z^{-2} + c_3 z^{-3} + c_4 z^{-4} + c_5 z^{-5} + c_6 z^{-6} + c_7 z^{-7} + c_8 z^{-8} + c_9 z^{-9} \quad (2)$$

Syntax:

```
linear <func name> <c0> <c1> <c2> <c3>
                         <c4> <c5> <c6> <c7> <c8> <c9>
```

where `func name` is `n0`, `n1`, `n2`, `n3` or `d`, and the cx's are the ten coefficients. The coefficients are floating-point numbers. All the calculations are carried out in floating-point single precision; therefore rounding errors might compromise stability if the poles are too close to the unit circle in the z plane. If you expect a controller to be stable and it is not, it is likely a rounding problem and, hence, you should try moving poles a little.
With no parameters, the command simply prints out the five functions n_0, n_1, n_2, n_3 and d.

Examples:

`linear n1 1 -0.3 4 7 1 0 0 0 0 0`
Sets $n_1(z) = 1 - 0.3z^{-1} + 4z^{-2} + 7z^{-3} + z^{-4}$.

`linear d 64 -62 0 0 0 0 0 0 0 0`
Sets $d(z) = 64 - 62z^{-1}$.

sinusoid

Sets the controller parameters in case the sinusoid controller is selected. The sinusoid controller is particular in that it is not a real controller. Rather, it is a sinusoidal signal generator that is independent of any of the four input values. The only two parameters that can be set are amplitude and frequency. This type of controller is useful for testing purposes.

Syntax:

`sinusoid <ampl. (volt)> <freq. (Hz)>`

`<ampl>` and `<freq>` are floating-point numbers.
With no parameters, it simply prints the current amplitude and frequency settings.

Examples:

`sinusoid`
Prints the current amplitude and the current frequency.

`sinusoid 1.1 0.75`
Instructs the controller to send a 0.75Hz, 1.1V-amplitude sinusoid to the Out output. The Out signal will then be amplified and shifted normally, according to the values set via the out command.

4 Understanding the Fuzzy Controller Features

Designing a fuzzy controller requires the definition of several aspects:

1. the input membership functions;
2. the output membership functions;
3. the control rules;
4. algorithms concerning fuzzyfication and de-fuzzyfication.

It is important to regard some of these aspects as fixed in order to allow an easier sensitivity study. In this section, item 4 is selected according to the previous experience of the authors with a fuzzy chip from STMicroelectronics [6, 10]. The same is done for item 1, with the only difference that here the dimensionless abscissa is 12-bit and, therefore, its range is from -2048 to 2047. On this axis it is then convenient to mount the membership function as much regularly as possible. Figure 5 gives an example for a total of five membership functions on the abscissa range. The link of the digital abscissa with the real values of the associated physical quantity is obtained by introducing a weight w. According to early studies, its value should scale the range of the digital abscissa to the range of the physical variable observed during an uncontrolled test. This would mean an undesired dependence of the controller on the excitation. By contrast, ongoing investigations suggest that the modulus

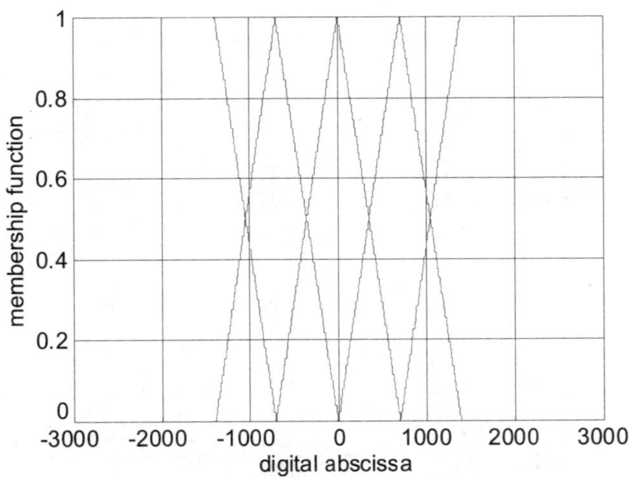

Fig. 5. Five membership functions on the digital abscissa range (-2048,2047)

of the more significant input achieves, in the controlled case, the value where the intermediate positive membership function reaches its peak. But the ratio between different input variables depends on the system and affects stability quite heavily [1].

For the algorithm adopted in this paper, the membership functions of the output are not explicitly required. It is sufficient to assign a crisp value to any of the attribute which characterize the output. For instance, for 7 attributes, one has to give a vector of values, **O**, of size 7.

With 5 membership functions for two inputs (for instance a couple of velocities) and 7 crisp values for the output (the control driving signal), three examples of rules are given in matrix form, in Tables I, II and III. For the sake of simplicity the usual labels zero (ZE), positive (PO) and negative (NE) are used, with PL (NL) meaning positive (negative) large, and XPL (XNL) positive (negative) extra-large.

The two input quantities are so that they require opposite counteractions. When one of the two quantities is null, the counteraction required spans from positive (negative) extra large to negative (positive) extra large. Along the main diagonal no counteraction is required.

Table 1 uses the values PL and NL of the output just for a single input combination. This creates significant valleys (peaks) between the central part of the response and the saturation zones (upper and lower plateau).

Figure 6 gives the graphical illustration of the control law generated by the rules of Table 1, for two different definitions of the crisp values positive (negative) and positive (negative) large, i.e., PO (NE) and PL (NL). The higher the crisp values adopted, the smoother the valley (peak) results.

Table 2 uses PL and NL all around the saturation regions. Figure 7 shows that this results in much smoother valleys (peaks) all around this region.

Table 1. Fuzzy control rules: example A

	NL	**NE**	**ZE**	**PO**	**PL**
NL	ZE	NE	XNL	NE	XNL
NE	PO	ZE	NE	NL	NE
ZE	XPL	PO	ZE	NE	XNL
PO	PO	PL	PO	ZE	NE
PL	XPL	PO	XPL	PO	ZE

Table 2. Fuzzy control rules: example B

	NL	**NE**	**ZE**	**PO**	**PL**
NL	ZE	NE	XNL	NL	XNL
NE	PO	ZE	NE	NL	NL
ZE	XPL	PO	ZE	NE	XNL
PO	PL	PL	PO	ZE	NE
PL	XPL	PL	XPL	PO	ZE

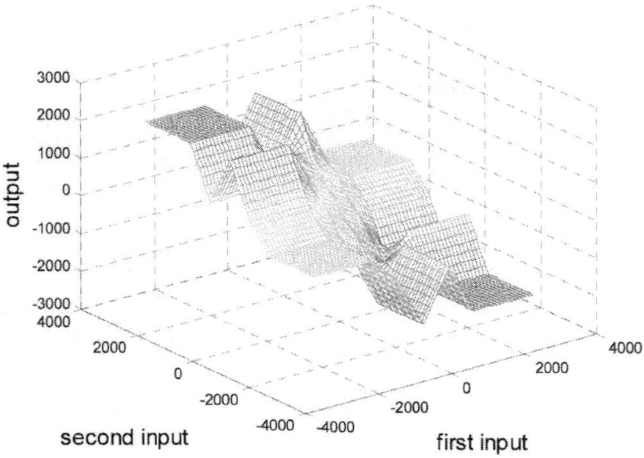

a)

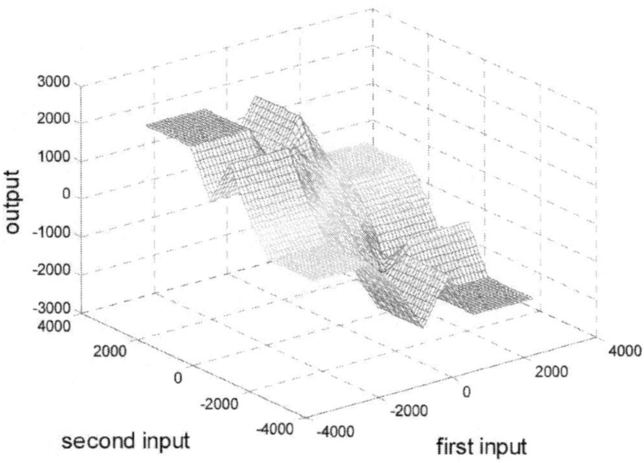

b)

Fig. 6. Fuzzy controller graphical representation: a) with intermediate crisp values 682, 1365; b) with intermediate crisp values 825, 1650

Table 3. Fuzzy control rules: example C

	NL	**NE**	**ZE**	**PO**	**PL**
NL	ZE	NE	XNL	NL	XNL
NE	PO	ZE	NE	ZE	NL
ZE	XPL	PO	ZE	NE	XNL
PO	PL	ZE	PO	ZE	NE
PL	XPL	PL	XPL	PO	ZE

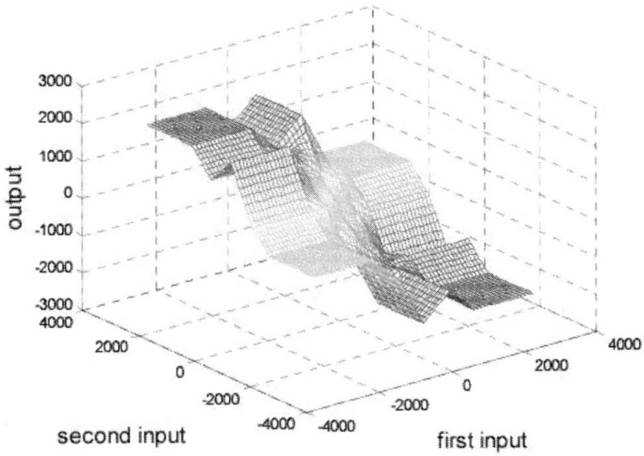

Fig. 7. Fuzzy controller graphical representation for example B (Table 2) with intermediate crisp values 682, 1365

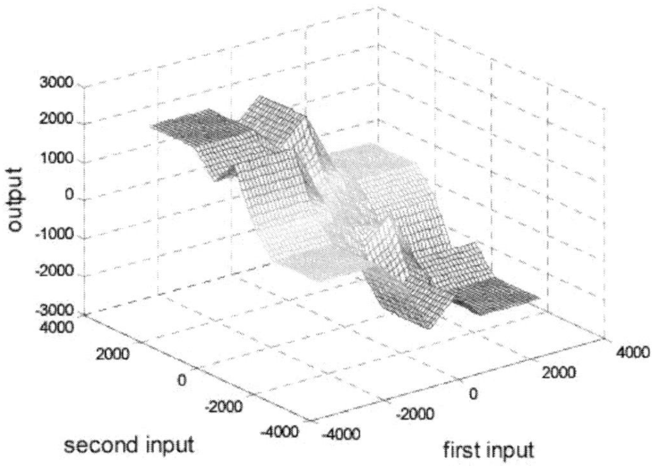

Fig. 8. Fuzzy controller graphical representation for example C (Table 3) with intermediate crisp values 682, 1365

Table 3 enlarges the area of no counteraction (starting from Table 2) and Figure 8 emphasizes how this set of rules realizes a sort of basin of no action for central values of the input variables.

Table 4 tries to summarize the free parameters in the design of a fuzzy controller with two input variables and one output. Each parameter is associated with a selection guide.

Table 4. Synthesis of the design parameters for a fuzzy controller

Parameter	Description	Selection process
w_1	Weight of the main input variable	The peak modulus of the controlled variable must reach the centroid of the intermediate positive membership function
w_2/w_1	Ratio of the input variable weights	System dependent: may affect stability
G	Gain factor	It should be equal to one for a successful design.
O	Vector of output crisp values	The null and peak values are given. The intermediate values can be selected to improve the control law shape for given rules.
RULES	IF-THEN relations between input variables and output.	They define the shape of the actual control law. Its general aspect can be given, but local details are adjusted by modifying the rules.

5 Conceiving a Wireless Implementation

One of the most challenging aspects of the current research activity concerns how to make the whole system wireless [11]. To this purpose, two choices are available. It must be decided whether to embrace an existing communication standard or to develop a new one. This dilemma stems from the idea that, although a proprietary protocol may be the best option available, as any desired feature can be stuffed in it, at the same time adopting an existing standard yields dramatically lower costs and development times. The real point is whether it is possible to find a standard that meets all the specifications.

In order to develop a completely new wireless communication protocol, many issues have to be taken into account. Perhaps the most difficult one, although not strictly of a technical nature, is what frequency band must be used. This depends on existing national regulations as well as on the desired values of range, power and channel capacity. Clearly, studying all these issues requires time and a well-defined set of specifications. This point is trickier than it might seem, because in several cases precise specifications are not known.

These considerations convinced the authors that, at least in the case under investigation, it is advisable to leave the wireless design work to the specialists and to focus attention on what must be done to complete the implementation. Therefore, the policy is to embrace an existing communication standard and to adopt off-the-shelf devices for wireless communication.

Several wireless standards are available on the market [5] but most of them have been developed for computer networks, for the Internet or for similar consumer applications. For example, Bluetooth is a low-cost, low-speed, short-range communication protocol that is intended mostly for interconnecting computer peripherals. Another example is the IEEE 802.11 standard, which allows more demanding networks to be implemented. Unfortunately, these two standards are too computer-friendly and, therefore, are not very easy to implement on a simple electronic board like the one discussed in this paper. It is too lightweight to be equipped with built-in networking capability. A much simpler solution was sought and two interesting devices were identified.

One is a full-fledged radio modem with an RS-232 interface and with the very interesting feature of allowing a few independent network sub-domains to be defined. This is very easy to use and requires no additional circuitry in order to be interfaced to the fuzzy controller board. It comes at a very reasonable price from Maxstream. In addition, it boasts a strikingly-long range, nearing 10 km in a line-of-sight situation. For the purposes of this paper, it would be perfect if it did not have the quite nasty shortcoming of low speed. The nominal speed is, in fact, 9.6 kbps. A higher speed option is available that enables full-duplex communication at 19.2 kbps, but this comes at the price of halving the achievable range.

The second device under consideration is a hybrid RF transceiver from Aurel that has a maximum speed of 100 kbps and a range just short of 1 km. As usually, speed has a cost in terms of range. This solution is more difficult to implement than the first one because additional circuitry is needed in order to interface it with the fuzzy controller module. However, it is so interesting as to deserve some investigation.

A further third solution that would be fairly easy to implement would be the GPRS protocol. There is a cheap module from Telit that acts as a modem with a serial interface and provides access to the Internet via the GPRS protocol. In this case the speed would be 57.6 kbps at most, but there would be no range limitations at all.

It is worth noting that serious reliability issues arise if these communication protocols are used for critical applications such as bridge control. One thing is monitoring a structure, another one is controlling it. Control loops pose very strict requirements on transmission latency and delay, while the problem of monitoring is much more relaxed. As of today, latency and delay are not guaranteed over the Internet. It is possible to request given levels of quality of service (QoS), but it does not seem very wise to have the structural stability of a bridge depend on a protocol that was not designed with real-time requirements in mind.

For these reasons, the two simpler protocols provided by the first two devices mentioned (i.e., the Maxstream radio modem and the Aurel hybrid RF transceiver) can offer, as of today, a more reliable solution. In both cases, the resulting sensor network is limited in space (while in the Internet case it would not) and, therefore, less prone to delays and network congestion. Also, steering clear from the Internet traffic, it should be more immune to tampering from potential attackers.

As anticipated before, designing a solution around the Maxstream radio modems requires no additional hardware at all. It is sufficient to plug in a radio

a) b) c)

Fig. 9. Some transceivers considered for the wireless extension. a) Stand alone radio modem (19.2 kbps) b) RF transceiver (100 kbps) c) GSM/GPRS module (up to 57.6 kbps)

modem to our fuzzy module and that will give it wireless capability. Only lesser software changes are needed in order to enable communication between two fuzzy modules, but that is independent of the communication protocol used, be it wireless or not. Figure 9 shows the three devices discussed above.

The following architecture was implemented in the authors' laboratory. A fuzzy module is used for signal acquisition from the accelerometers. Its fuzzy capabilities are not used in this case. It just reads from the accelerometers and routes the data through the radio modem to the second fuzzy module. The second fuzzy module receives the data and performs the actual work. Based on the accelerations received, it operates an active mass damper actuator and strives to keep a structural specimen from failure.

An aspect that could be of some concern when dealing with wireless systems is the possibility of interference. When a disturbing transmitter enters a wireless network, it is quite normal to expect that something might cease working properly. While this is certainly true in the case of AM or FM transmitters or, more in general, of traditional radio transmitters, the type of modulation used in the two solutions described above is inherently more robust to this kind of problems. Rather than using a separate channel for every transmitter, which would sound very intuitive, the spread-spectrum modulation uses the opposite approach. All the transmitters share a common frequency band. This band is divided into many independent sub-channels. When a device needs to transmit data, it picks one sub-channel and uses it for a very short time. Then it hops to another sub-channel. This way, interferences are not a serious problem because interferences are accepted. They are part of the game. In case a sub-channel is busy and a collision occurs, it is sufficient to hop to a new sub-channel and retransmit. Clearly, frequency hopping lowers the probability that two transmitters may interfere with each other and interference only causes a limited reduction of the connection speed. This can only become a problem when nearing the network congestion limit. An advantage of the radio modem we used is that all the frequency hopping and collision detecting logic is automatically handled and, therefore, there is no extra care to pay to the problem.

6 Results of Laboratory Testing

This section reports the results of some laboratory tests conducted on the frame of Figure 10, mounting on the top an active mass damper driven by the fuzzy board of Figure 2. The input variables to the board are the acceleration of the three storeys, as well as that of the shaking table. All the experiments were conducted by exciting the table with a white-noise acceleration. The peak amplitude of the piston stroke is 4.9 mm when the system is tested without control. Then tests with the maximum stroke multiplied by an integer factor from 1 to 4 were also carried out.

The input variables for the fuzzy controller are the velocities of the second and third floor. The velocities are absolute velocities: tests with relative (to the base) velocities resulted too biased by the current value of the base velocity. The weights selected are 12.5 for the third floor (main variable) and -3 for the second floor. The gain is set to one. The controller of Table 1 is adopted, with the intermediate values at 682, 1365 (Figure 6a), case A1; 750, 1500, case A2; 825, 1650 (Figure 6b), case A3.

Fig. 10. The experimental environment: three-storey steel frame with active mass damper on the top

Table 5. The tests reported in this paper. The stroke span factor should be multiplied by 4.9 mm to provide the stroke span

Number	Controller	Stroke span factor	Index
1	Uncontrolled	1	364.09
2	A1	1	176.51
3	A3	1	150.97
4	A1	2	333.78
5	A2	2	313.02
6	A3	2	288.01
7	A1	3	503.86
8	A2	3	444.76
9	A3	3	435.01
10	A2	4	627.80
11	A3	4	569.13

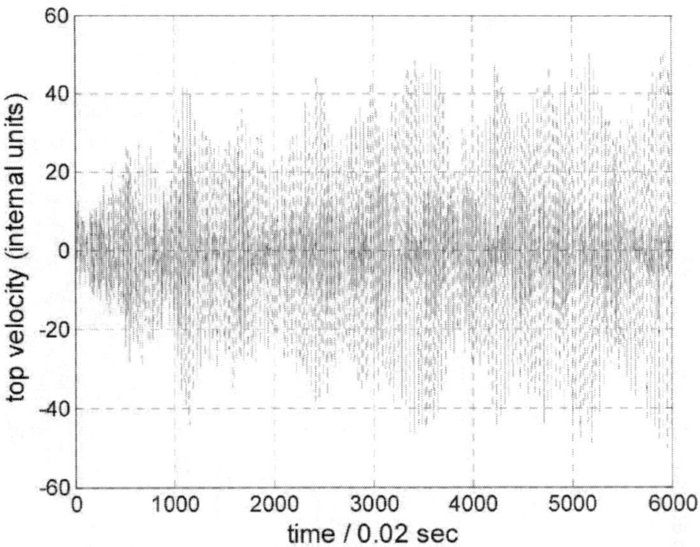

Fig. 11. Span stroke factor equal to one. Comparison of the top velocity time history in the uncontrolled case (dotted line) and in the A1 controlled case (solid line)

Table 5 provides a synthesis of the tests reported in this section. The column named index is defined as the mean of the 6 external histogram bars of the top velocity statistical distribution after removing the inner four bars. It gives a measure of the excursions of the response far from the rest position.

Figure 11 mainly shows the amount of response improvement the control allows in a case where the uncontrolled test was feasible (stroke span factor equal to 1). Figure 12 emphasizes the different behaviour of controllers A1 and A3.

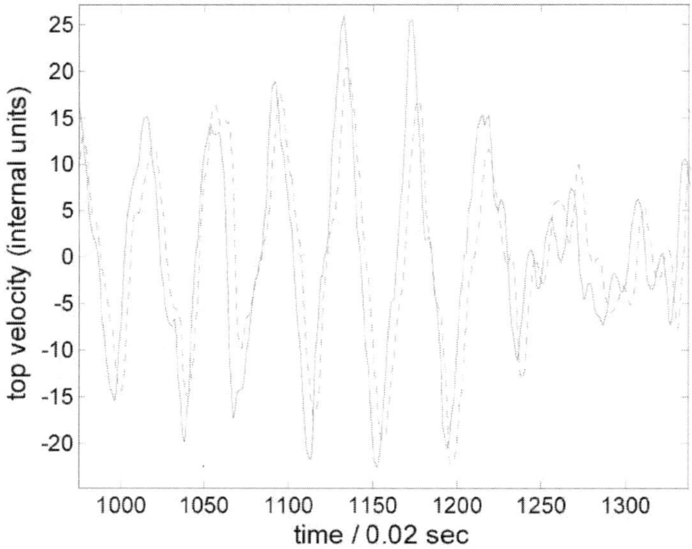

Fig. 12. Span stroke factor equal to one. Comparison of the top velocity time history in the A1 controlled case (solid line) and in the A3 controlled case (dotted line)

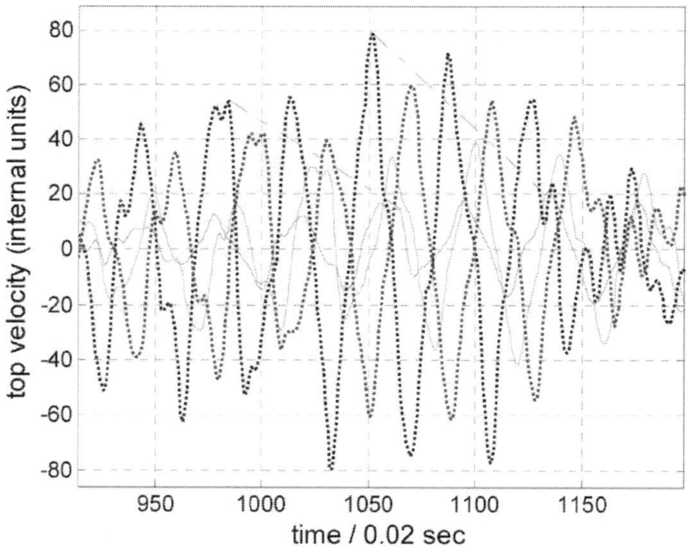

Fig. 13. Controlled top velocity by the A3 controller for different values of the stroke span factor (from 1 to 4)

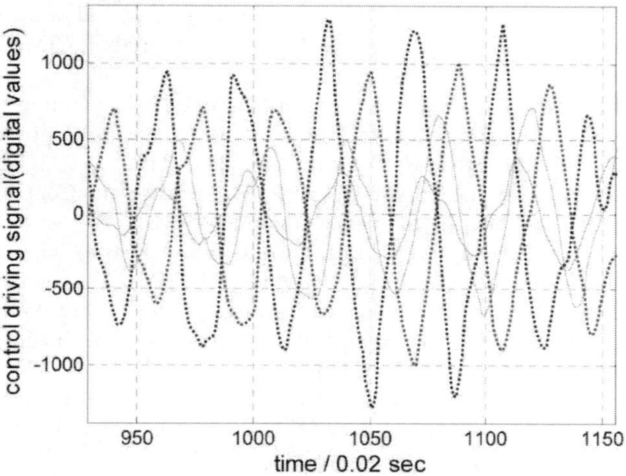

Fig. 14. Control signal (in digital value) generated by the A3 controller for different values of the stroke span factor (from 1 to 4)

Figure 13 shows the different response for different values of the stroke span factor (from 1 to 4), while Figure 14 provides the same comparison in terms of output signal.

7 Conclusions

The main contribution of this paper covers the implementation of a board mounting a microprocessor programmed as a fuzzy controller. The software shell is illustrated and the architecture for a wireless solution is investigated.

Some laboratory tests are conducted to validate the whole procedure and to identify areas where new improvements should be pursued.

Acknowledgement. This work is funded by the Italian Space Agency (ASI) (Grant number ASI IR 10-2002).

References

1. Casciati F (1997) Checking the Stability of a Fuzzy Controller for Nonlinear Structures, Microcomputers in Civil Engineering 12:205-215
2. Casciati F (1998) Scale Laboratory Testing towards Actual Actively Controlled Structural Systems. Proc. of the 2nd World Conference on Structural Control, T. Kobori et al. (eds.), John Wiley, Chichester, UK, pp. 41-50
3. Casciati F (ed.) (2003) Proceedings 3rd World Conference on Structural Control, John Wiley & Sons, Chichester

4. Casciati F, Magonette G (1999) Testing Facilities and Laboratory Validation, in: Advances in Structural Control, J. Rodellar et al. (eds.), CIMNE, Barcelona, pp. 241-266
5. Casciati F, Faravelli L, Borghetti F, Fornasari A (2003) Tuning the frequency band of a wireless sensor network, in: Fu-Kuo Chang (ed.) Structural Health Monitoring 2003, DEStech Publications, pp. 1185-1192
6. Casciati F, Faravelli L, Torelli G (1999) A Fuzzy Chip Controller for Civil Applications, Nonlinear Dynamics 20(1):pp.85-98
7. Faravelli L (1999) Fuzzy Logic, Fuzzy Control and Fuzzy Chips, in: Advances in Structural Control, J. Rodellar et al. (eds.), CIMNE, Barcelona, pp. 226-240
8. Faravelli L (2002) Innovative Control Technologies for Vibration Sensitive Civil Engineering Structures in Europe, National Workshop on Future Sensing Systems, Lake Tahoe, Granlibakken, USA
9. Faravelli L, Rossi R (2000) Fuzzy Chip Controller Implementation, in: Casciati F and Magonette G (eds.) Proc. 3rd International Workshop on Structural Control, World Scientifics, Singapore, pp. 201-213
10. Faravelli L, Rossi R (2002) Adaptive Fuzzy Control: Theory versus Implementation, J. Structural Control 9:59-73
11. Faravelli L, Rossi R (2003) Wireless Communication between Sensor/device Stations, Proceedings ICES02, Balkema, Rome
12. Faravelli L, Spencer JB (eds.) (2003) Proc. of the US-Europe Workshop on Sensors and Smart Structures Technology, John Wiley & Sons, Chichester, UK
13. Wong F, Chou K, Yao J (1999) Civil Engineering Including Earthquake Engineering, in: H. Zimmermann (ed.), Practical Applications of Fuzzy Technologies, Kluwer Academic Press, Dordrecht

Semi-Active Friction Damping of Flexible Lightweight Structures

Lothar Gaul, Hans Albrecht, and Jan Wirnitzer

Institute A of Mechanics, University Stuttgart, Germany
gaul@mecha.uni-stuttgart.de

Summary. The present approach for vibration suppression of flexible structures is based on friction damping in semi-active joints. At optimal locations conventional rigid connections of a large truss structure are replaced by semi-active friction joints. Two different concepts for the control of the normal forces in the friction interfaces are implemented. In the first approach each semi-active joint has its own local feedback controller, whereas the second concept uses a global, clipped-optimal controller. Simulation results of a 10-bay truss structure show the potential of the proposed semi-active concept.

1 Introduction

Large space structures (LSS) are commonly designed as very flexible and lightweight truss structures of big size. Due to their small damping and the stringent positioning accuracy, many missions involving antennas or optical interferometers require vibration suppression. Numerous works have been published on active vibration suppression. Usually piezoelectric stacks are used as devices for actuators because of their light weight, high force and minimum power consumption. Although an active approach is very attractive, actively controlled systems may cause spillover instability. Passive approaches for damping enhancement involving visco-elastic materials or preloaded backlashes [1] may be desirable because of their simplicity and low cost. The present approach is based on friction damping in the joint connections of a truss structure [2]. Dry friction caused by slip in the interface of connected parts turns out to provide a significant amount of energy dissipation. The original truss nodes are modified such that relative slip between the truss member's end connector and the truss node is provided.

However, passive approaches for vibration suppression are by far not as effective as active ones. Passive friction damping has several disadvantages. When the amplitude of vibration decreases below a certain level sticking occurs and energy is no longer dissipated. In addition, if the stiction force is quite high the connected parts can stick and thus static shape accuracy cannot be assured. To overcome these drawbacks, the friction force in the joint connection is controlled by varying the normal force in the contact interface using a piezoelectric stack [3]. Since a passive device is actively controlled, this approach is called semi-active. The resulting semi-active vibration suppression system is always stable, at least in the Lyapunov sense, because of the dissipative nature of friction. Another appeal of semi-active control is that performance levels of active control can be achieved with a fraction of input power. Furthermore, this concept can easily be employed without increasing the structural weight.

The paper is structured as follows. It starts with the numerical model of the adaptive structure consisting of the truss structure which is considered as a linear subsystem and the nonlinear semi-active joints which exert state-dependent friction forces on the linear subsystem. The parameters of the nonlinear friction model must be identified from measurements conducted with an isolated joint. Based on the open-loop state space model of the linear subsystem, modal reduction is performed using controllability and observability gramians. To improve the fidelity of the reduced model for lower frequencies, the modal subspace is augmented with Krylov vectors.
At optimal locations conventional connections are replaced by semi-active friction joints and two different control concepts for the semi-active vibration suppression are introduced: local feedback control and clipped-optimal control. The simulation results for a 10-bay truss structure presented at the end reveal the potential of the presented semi-active approach.

2 Adaptive Truss Structure

The 10-bay truss structure depicted in Fig. 1 is made of Meroform-M12 construction system consisting of aluminum tubes connected by steel nodes. The Meroform components are shown in more detail in Fig. 2. Each tube has a 22 mm outer diameter and is fitted with a screw end connector. The horizontal and vertical distance between the center of the truss nodes is 1 m. In accordance with the Shuttle Radar Transmission Mission (SRTM)[1] where a similar truss structure has been used as a cantilever to carry one of the two radar equipments, the mast is clamped at one end. At the free end a mass of 0.5 kg is attached to each of the five truss nodes.

For the implementation of the semi-active damping approach a modified joint connection has been designed, which allows relative motion between the end connector of a truss member and the truss node with the remaining members attached to it. The rigid Meroform nodes can be replaced by this adaptive joint. As shown in Fig. 3, two types of joints are considered, each

[1] For details visit the websites `http://www.jpl.nasa.gov/srtm/` and `http://www.aec-able.com/corporate/srtm.htm`

Fig. 1. Truss structure

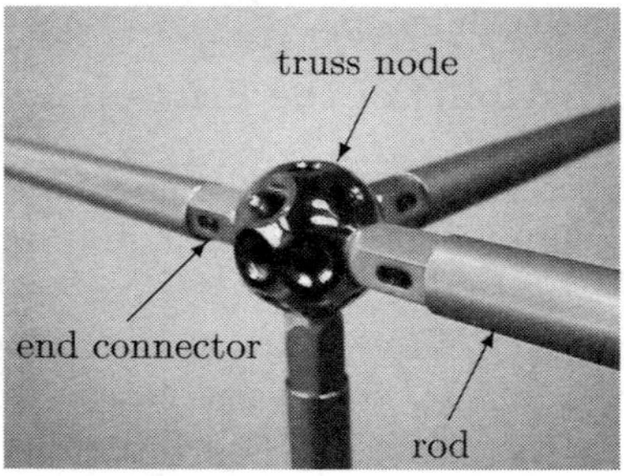

Fig. 2. Meroform components

with one single degree-of-freedom. It is the relative displacement along the longitudinal axis of the connected rod (type A) and relative rotation about an axis perpendicular to the longitudinal axis (type B).

A sketch of the rotational prototype (type B) is shown in Fig. 4. The semi-active joint employs the dissipative character of dry friction caused by interfacial slip in the contact area[2]. By applying voltage to the piezoelectrical stack actuator, the pressure in the contact area can be altered. The normal force is approximately proportional to the applied voltage. The preload of the stack can be adjusted by a screw.

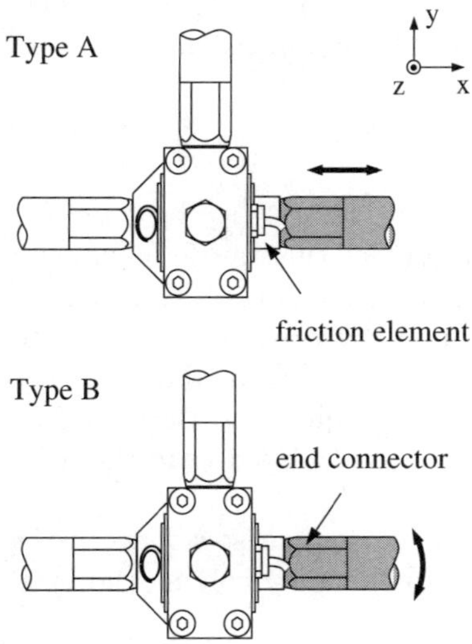

Fig. 3. Types of joints

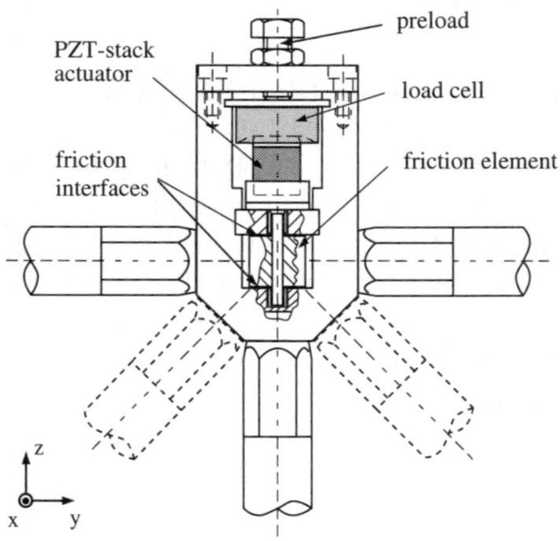

Fig. 4. Adaptive joint (type B)

[2] L. Gaul: German Patent "Smart Joint", Number 19702518 C2 (1997)

3 Numerical Model

3.1 Model of the Friction Joint

For the modeling of the nonlinear behaviour in the friction joints the so called LuGre-model [4] has been proposed [3] which is capable of simulating relevant friction phenomena [5], such as presliding displacement, stick-slip and the Stribeck effect. It describes the friction interface as a contact between bristles. The internal state variable φ representing the average deflection of the bristles is governed by the first order differential equation

$$\dot{\varphi} = f(v, \varphi) = v - \sigma_0 g(v)\, \varphi \qquad \text{with} \qquad g(v) = \frac{|\, v \,|}{F_c + F_d e^{-(v/v_s)^2}} \tag{1}$$

where v is the relative sliding velocity in the contact. The friction force transmitted in the joint

$$F_R = F_N \underbrace{(\sigma_0\, \varphi + \sigma_1\, \dot{\varphi} + \sigma_2\, v)}_{=:\mu\,(v,\,\varphi)} \tag{2}$$

depends on the normal force F_N and the dynamic friction coefficient $\mu(v, \varphi)$. In case of a rotational joint (type B), a radius r is introduced in (2) to obtain a friction moment, i.e., $M_R = r\, F_R$, where v represents the relative angular velocity. In the following, for both types the generalised force u_R and the generalised sliding velocity v will be used. The model parameters $\sigma_{0_k}, \sigma_{1_k}, \sigma_{2_k}, F_{c_k}, F_{d_k}$ and v_{s_k} are all positive and have dimensions consistent with the dimension of v.

3.2 Parameter Identification

In order to identify the parameters of the LuGre-model, the transfer behaviour of an isolated friction joint is measured. The joint is supported on a foundation and the friction element (see Fig. 4) is harmonically excited by a shaker. The sliding velocity and the excitation force are recorded for different normal forces and different excitation amplitudes and frequencies. Based on the known inertia the transmitted friction force can be calculated from the measured excitation force.

The model parameters are determined from the comparison between the simulated response and the measurements. This is done by solving the optimization problem

$$\min_{\boldsymbol{p}} \mathcal{J}(\boldsymbol{p}) = \frac{1}{2} \sum_{k=0}^{K} \left(u_R(t_k) - \hat{u}_R(v(t_k), F_N, \boldsymbol{p})\right)^2 \tag{3}$$

with the vector $\boldsymbol{p}$ containing the parameters to be identified, the measured friction force $u_R(t_k)$ and the simulated friction force $\hat{u}_R(v(t_k), F_N, \boldsymbol{p})$, which is obtained from (2) and integration of (1). For calculation of $\hat{u}_R$ the measured sliding velocity $v(t_k)$ and normal force F_N which is kept constant during the recording period are used.

From the results of a fuzzy sensitivity analysis [6] it is known, that the individual parameters are responsible for different behaviour characteristics. Micro-slip behaviour is mostly affected by the parameters $\boldsymbol{p}_1 = [\sigma_0 \ \sigma_1]$. During macro-slip, where sliding in the entire contact area occurs, especially the parameters $\boldsymbol{p}_2 = [F_c \ F_d \ v_s \ \sigma_2]$ are of importance. Macro-slip and micro-slip behaviour can be generated by choosing an appropriate combination of excitation amplitude and normal force which has an influence on the break-away force.

To efficiently solve the optimization problem in (3), the identification is carried out in two steps. First, the parameters $\boldsymbol{p}_1$ are idenitified from responses measured during micro-slip regime. The remaining parameters $\boldsymbol{p}_2$ are obtained from measurements where the joint is operated in stick-slip behaviour. For both optimizations, data from several measurements with different excitations and normal forces are used. Because of the nonlinear objective function (3) a gradient-based optimization algorithm will only give local minima. Therefore, it is important to start the optimization with reasonable initial values for $\boldsymbol{p}_1$ and $\boldsymbol{p}_2$.

The identified parameters are listed in Table 1 for both types of joint. The present joint design turned out to have quite small parameters σ_1 und σ_2 which have almost no influence on the dynamical behaviour. A fact that is confirmed by the results of the fuzzy sensitivity analysis [6]. Therefore, it is hard to identify there actual values. As an example, one of the measured hysteresis loops for type B is compared to the predicted loop in Fig. 5.

Table 1. Identified friction parameters

Typ	σ_0	σ_1	σ_2	F_c	F_d	v_s	r
A	$1.42 \cdot 10^5$ 1/m	0.02 s/m	0.01 s/m	0.61	0.14	0.01 m/s	–
B	$6.5 \cdot 10^4$ 1/rad	0.02 s/rad	0.01 s/rad	0.49	0.08	0.02 rad/s	0.011 m

3.3 FE-Model of the Truss Structure

The FE discretization of the truss connections is depicted in Fig. 6. All tubes and end connectors are discretized by cubic Euler-Bernoulli beams with two nodes, each having six degrees of freedom. The Meroform nodes are considered as perfectly rigid and the rigid portions are modelled by geometric constraints $\boldsymbol{u}_m = \boldsymbol{S}_{nm}\boldsymbol{u}_n$ between the tied dofs $\boldsymbol{u}_m$ and the independent dofs $\boldsymbol{u}_n$. The mass and inertia of the Meroform connections are lumped at the centers of each truss node. The kinematics of the friction joints are described by

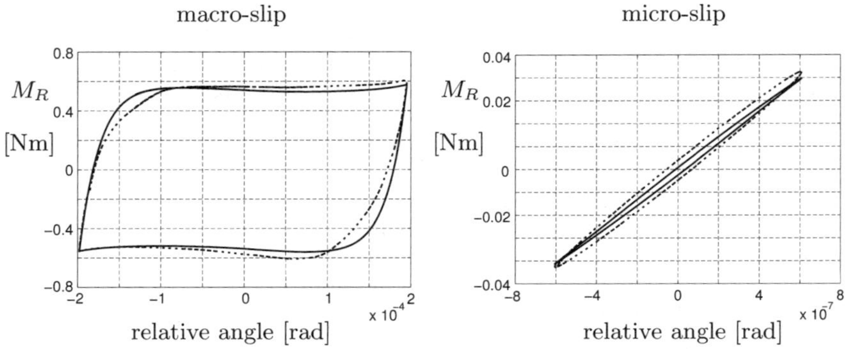

Fig. 5. Hysteresis loops for type B: simulation (——), experiment (– –)

constraint matrices, as well, which couple the degrees of freedom of the corresponding nodes according to the type of joint. By assembling all constraint matrices $\boldsymbol{S}_{nm}$ into matrix $\boldsymbol{S}$, all dependent degrees of freedom of the original finite element model with mass matrix $\bar{\boldsymbol{M}}$, damping matrix $\bar{\boldsymbol{D}}$, stiffness matrix $\bar{\boldsymbol{K}}$ and the displacement vector $\bar{\boldsymbol{z}}$ can be eliminated as follows

$$\boldsymbol{S}^T\bar{\boldsymbol{M}}\boldsymbol{S}\ddot{\bar{\boldsymbol{z}}} + \boldsymbol{S}^T\bar{\boldsymbol{D}}\boldsymbol{S}\dot{\bar{\boldsymbol{z}}} + \boldsymbol{S}^T\bar{\boldsymbol{K}}\boldsymbol{S}\bar{\boldsymbol{z}} = \boldsymbol{S}^T\bar{\boldsymbol{E}}\,\boldsymbol{u}_R\,(\boldsymbol{v},\boldsymbol{\varphi}) + \boldsymbol{S}^T\bar{\boldsymbol{F}}\boldsymbol{w}\;. \tag{4}$$

The resulting system of differential equations is given by

$$\boldsymbol{M}\ddot{\boldsymbol{z}} + \boldsymbol{D}\dot{\boldsymbol{z}} + \boldsymbol{K}\boldsymbol{z} = \boldsymbol{E}\,\boldsymbol{u}_R\,(\boldsymbol{v},\boldsymbol{\varphi}) + \boldsymbol{F}\boldsymbol{w} \tag{5}$$

with the vector $\boldsymbol{z}$ containing only the independent generalised displacements z_j $(j = 1,\ldots,N)$. The friction forces $\boldsymbol{u}_R(\boldsymbol{v},\varphi)$ acting in the friction joints depend on the sliding velocities $\boldsymbol{v} = \boldsymbol{E}^T\dot{\boldsymbol{z}}$ and the bristle deflections $\boldsymbol{\varphi}$. Vector $\boldsymbol{w}$ represents disturbances.

Using the modal transformation

$$\boldsymbol{z}(t) = \boldsymbol{\Phi}\boldsymbol{\eta}(t) \tag{6}$$

with the modal matrix $\boldsymbol{\Phi} = [\boldsymbol{\phi}_1,\ldots,\boldsymbol{\phi}_n]$ (including $n < N$ 'significant' eigenmodes $\boldsymbol{\phi}_i$) obtained from the undamped eigenvalue problem

$$\boldsymbol{K}\boldsymbol{\phi}_i = \omega_i^2\boldsymbol{M}\boldsymbol{\phi}_i\;, \qquad i = 1,\ldots,n \tag{7}$$

and assuming proportional damping, (5) can be transformed to a set of differential equations

$$\ddot{\boldsymbol{\eta}} + \boldsymbol{\Xi}\dot{\boldsymbol{\eta}} + \boldsymbol{\Omega}\boldsymbol{\eta} = \boldsymbol{\Phi}^T \boldsymbol{E}\, \boldsymbol{u}_R(\boldsymbol{v}, \boldsymbol{\varphi}) + \boldsymbol{\Phi}^T \boldsymbol{F} \boldsymbol{w} \tag{8}$$

with the matrices $\boldsymbol{\Omega} = \mathrm{diag}\,(\omega_i^2)$, $\boldsymbol{\Xi} = \mathrm{diag}\,(2\xi_i\,\omega_i)$ and the vector $\boldsymbol{\eta}$ of modal coordinates η_i. The eigenfrequency and the modal damping coefficient of the i-th mode is denoted by ω_i and ξ_i, respectively.

The sliding velocities $\boldsymbol{v}$ and additional measurements $\boldsymbol{y}$ are given by

$$\boldsymbol{v} = \boldsymbol{E}^T \boldsymbol{\Phi}\, \dot{\boldsymbol{\eta}} \tag{9}$$

$$\boldsymbol{y} = \boldsymbol{V}\boldsymbol{\Phi}\boldsymbol{\eta} + \boldsymbol{W}\boldsymbol{\Phi}\dot{\boldsymbol{\eta}}\,. \tag{10}$$

It should be pointed out that the transformation matrix in (6) is composed of eigenvectors of the underlying linear undamped system. That means, the nonlinear state-dependent term in the excitation vector $\boldsymbol{u}_R$ is neglected and a truss structure with frictionless joint connections is considered, i.e. $F_{N_j} = 0$. That way, the structure of the linear subsystem can fully be employed for the upcoming model reduction. Such an approach is often used for systems with local nonlinearities [7].

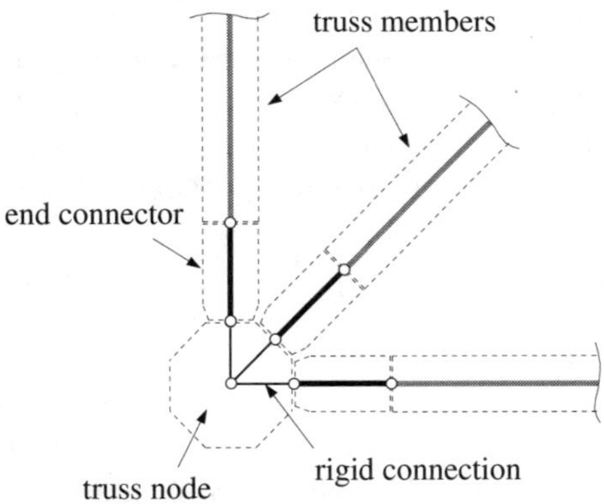

Fig. 6. FE discretization

Table 2. Eigenfrequencies in [Hz]

No	EMA	FEM	error [%]	MAC
1	4.49	4.495	0.18	0.79
2	4.60	4.595	0.19	0.88
3	20.65	20.60	0.24	0.97
4	22.07	22.25	0.82	0.95
5	23.61	23.76	0.64	0.93
7	24.72	25.08	1.46	0.93
8	43.47	43.73	0.60	0.95
12	46.83	46.90	0.15	0.98
17	49.45	49.52	0.14	0.72

Since not all of the parameters of the FE model are known a priori, they have to be identified by means of Modal Updating [8] using eigenfrequencies and mode shapes obtained from an Experimental Modal Analysis (EMA) which has been performed on the struture. The unknown parameters particularly are the axial, bending and torsional stiffness of the end connectors. Furthermore, the material properties of the aluminum members are not exactly known. Dimensions and masses of the components however are considered as given.

The eigenfrequencies calculated with the updated FE model are compared to the measured values in Table 2. The corresponding MAC values [8] of the mode shapes are listed as well. Note that only those modes are listed whose mode shapes could clearly be identified by the EMA. The extracted modal damping ratios $0.05\% \leq \xi_i \leq 0.5\%$ confirm that the present strutcture indeed is a weakly damped system. For the upcoming simulations $\xi_i = 0.1\%$ will be used for all modes $i = 1, \ldots, n$.

3.4 Model of the Adaptive Structure

By defining the state vector of the linear subsystem as

$$\boldsymbol{x} = \left[\boldsymbol{x}_1^T, \ldots, \boldsymbol{x}_n^T\right]^T \qquad \text{with} \qquad \boldsymbol{x}_i = \begin{bmatrix} \dot{\eta}_i \\ \omega_i \eta_i \end{bmatrix} \tag{11}$$

where the state $\boldsymbol{x}_i$ represents the i-th mode, the state space form of the friction damped truss structure is given by

$$\begin{bmatrix} \dot{\boldsymbol{x}} \\ \dot{\boldsymbol{\varphi}} \end{bmatrix} = \begin{bmatrix} \boldsymbol{A}\boldsymbol{x} \\ \boldsymbol{f}(\boldsymbol{v}, \boldsymbol{\varphi}) \end{bmatrix} + \begin{bmatrix} \boldsymbol{B} \\ \boldsymbol{0} \end{bmatrix} \boldsymbol{u}_R(\boldsymbol{v}, \boldsymbol{\varphi}) + \begin{bmatrix} \boldsymbol{H} \\ \boldsymbol{0} \end{bmatrix} \boldsymbol{w} \,. \tag{12}$$

The system matrix $\boldsymbol{A}$ as well as the matrices $\boldsymbol{B}$ and $\boldsymbol{H}$ can easily be deduced from (5) by taking the definition of the state vector $\boldsymbol{x}$ into account. The internal state variables φ_j $(j = 1, \ldots, l)$ of the l semi-active friction joints are arranged in $\boldsymbol{\varphi}$, the corresponding differential equations (1) in $\boldsymbol{f}(\boldsymbol{v}, \boldsymbol{\varphi})$. The sliding velocities are given by $\boldsymbol{v} = \boldsymbol{B}^T \boldsymbol{x}$. The actuator forces $\boldsymbol{u}_R$ are defined by (2) as $\boldsymbol{u}_{R_j} = F_{N_j}\, \mu_j(v_j, \varphi_j)$ where the normal forces f_{N_j} represent the controller output.

4 Model Reduction

Because of the high order of the system (12), a model reduction is required to obtain a low order model for the control design. This is done by significantly reducing the order of the linear subsystem and keeping the nonlinear part unchanged. Thus, the linear subsystem

$$\begin{bmatrix} \dot{\boldsymbol{x}}_d \\ \dot{\boldsymbol{x}}_{nd} \end{bmatrix} = \begin{bmatrix} \boldsymbol{A}_d & 0 \\ 0 & \boldsymbol{A}_{nd} \end{bmatrix} \begin{bmatrix} \boldsymbol{x}_d \\ \boldsymbol{x}_{nd} \end{bmatrix} + \begin{bmatrix} \boldsymbol{B}_d \\ \boldsymbol{B}_{nd} \end{bmatrix} \boldsymbol{u} + \begin{bmatrix} \boldsymbol{H}_d \\ \boldsymbol{H}_{nd} \end{bmatrix} \boldsymbol{w} \tag{13}$$

$$\boldsymbol{v} = \begin{bmatrix} \boldsymbol{B}_d^T \ \boldsymbol{B}_{nd}^T \end{bmatrix} \begin{bmatrix} \boldsymbol{x}_d \\ \boldsymbol{x}_{nd} \end{bmatrix} \tag{14}$$

$$\boldsymbol{y} = \begin{bmatrix} \boldsymbol{C}_d \ \boldsymbol{C}_{nd} \end{bmatrix} \begin{bmatrix} \boldsymbol{x}_d \\ \boldsymbol{x}_{nd} \end{bmatrix} \tag{15}$$

is considered in the following. It is partitioned into dominant states $\boldsymbol{x}_d$ which will be kept in the reduced model, and states $\boldsymbol{x}_{nd}$ which will be removed. In the following considerations, the nonlinear friction forces $\boldsymbol{u}_R$ are neglected and the vector $\boldsymbol{u}$ represents external (state-independent) actuator forces.

4.1 Balanced Reduction

The determination of dominant modes $\boldsymbol{x}_{d_i}$ is based on balanced reduction using controllability and observability gramians corresponding to the different inputs ($\boldsymbol{u}$, $\boldsymbol{w}$) and outputs ($\boldsymbol{v}$, $\boldsymbol{y}$) of the system. For an asymptotically stable system, the controllability gramian $\boldsymbol{W}^c$ and observability gramian $\boldsymbol{W}^o$ are symmetric and positive definite matrices which can be computed from the steady-state Lyapunov equations [9]

$$\boldsymbol{A}\boldsymbol{W}^c + \boldsymbol{W}^c\boldsymbol{A}^T + \boldsymbol{B}\boldsymbol{B}^T = \boldsymbol{0} \qquad \text{and} \qquad \boldsymbol{A}^T\boldsymbol{W}^o + \boldsymbol{W}^o\boldsymbol{A} + \boldsymbol{C}^T\boldsymbol{C} = \boldsymbol{0}\,. \quad (16)$$

Due to the particular form of the state space model with modal coordinates, (16) can be solved in closed form [10]. For structures with small damping and well separated eigenfrequencies[3] the gramians tend to diagonal matrices

$$\boldsymbol{W}^{c(o)} = \text{diag}\begin{pmatrix} w_i^{c(o)} & 0 \\ 0 & w_i^{c(o)} \end{pmatrix} \quad (17)$$

$$\text{with} \quad w_i^c = u_i\,\frac{\boldsymbol{e}_i\boldsymbol{e}_i^T}{4\xi_i\omega_i}\,, \qquad w_i^o = \frac{\omega_i^2\boldsymbol{w}_i^T\boldsymbol{w}_i + \boldsymbol{v}_i^T\boldsymbol{v}_i}{4\xi_i\omega_i^3}$$

where $i = 1,\ldots,n$. The vector $\boldsymbol{e}_i$ is the i-th row of $\boldsymbol{\Phi}^T\boldsymbol{E}$, vectors $\boldsymbol{v}_i$ and $\boldsymbol{w}_i$ corespond to the i-th column of $\boldsymbol{V\Phi}$ and $\boldsymbol{W\Phi}$, respectively. The i-th diagonal element w_i^c (w_i^o) is a quantitative measure for the controllability (observability) of the i-th state $\boldsymbol{x}_i = [\dot{\eta}_i, \omega_i\eta_i]^T$ which represents the i-th mode.

To identify modes which are relevant for the transfer behaviour of the input/output path from $\boldsymbol{u}_R$ to $\boldsymbol{v}$, one has to check the corresponding Hankel singular values [11]

$$\gamma_{uv_i} = \sqrt{w_{u_i}^c\, w_{v_i}^o} = \frac{\boldsymbol{e}_i\boldsymbol{e}_i^T}{4\xi_i\omega_i} \quad (18)$$

defined by means of the eigenvalues of the controllability and observability gramians for the input $\boldsymbol{u}$ and output $\boldsymbol{v}$, i. e., $w_{u_i}^c$ and $w_{v_i}^o$ respectively. Modes with large Hankel singular values γ_{uv_i} should be included into the reduced model in order to guarantee a good approximation of the original system. Those modes will mainly be excited by the friction forces. And they will contribute to the sliding velocities $\boldsymbol{v}$ occuring in the friction joints, as well.

In addition, all modes which have an impact on the system measurements $\boldsymbol{y}$ need to be added to the reduced model. They are selected according to the eigenvalues $w_{y_i}^o$ of the observability gramian corresponding to the output $\boldsymbol{y}$. When the disturbance input matrix $\boldsymbol{H}$ is known, it can be taken into account as well.

[3] Even for truss structures with their quite high modal density, this approximation is valid as long as the damping ratios are small, i.e., $\xi_i \ll 1$ [10, 11].

4.2 Krylov Vectors

After the dominant states are chosen, the reduced model of the linear subsystem is given by

$$\dot{\boldsymbol{x}}_d = \boldsymbol{A}_d \boldsymbol{x}_d + \boldsymbol{B}_d \tilde{\boldsymbol{u}}_R(\tilde{\boldsymbol{v}}, \boldsymbol{\varphi}) + \boldsymbol{H}_d \boldsymbol{w} \tag{19}$$

$$\tilde{\boldsymbol{v}} = \boldsymbol{B}_d^T \boldsymbol{x}_d = \boldsymbol{E}^T \boldsymbol{\Phi}_d \dot{\boldsymbol{\eta}}_d \tag{20}$$

$$\tilde{\boldsymbol{y}} = \boldsymbol{C}_d \boldsymbol{x}_d = \boldsymbol{V} \boldsymbol{\Phi}_d \boldsymbol{\eta}_d + \boldsymbol{W} \boldsymbol{\Phi}_d \dot{\boldsymbol{\eta}}_d \, . \tag{21}$$

Now, focus on the input/output path from $\boldsymbol{u}_R$ to $\boldsymbol{v}$ which is the most critical since the prediction of the local nonlinearities is effected by the approximation quality of this path. As a result of neglecting the part $\boldsymbol{B}_{nd}^T \boldsymbol{x}_{nd}$ in (20), the approximated sliding velocity $\tilde{\boldsymbol{v}}$ is erroneous. This has a negative effect on the prediction of the resulting friction forces $\tilde{\boldsymbol{u}}_R(\tilde{\boldsymbol{v}}, \boldsymbol{\varphi})$. The approximation error would amplify when simulating the closed-loop system. Therefore, it is important to somehow compensate for the deleted modes.

For this purpose the transfer matrix $\boldsymbol{G}^{uv}(s)$ from $\boldsymbol{u}$ to $\boldsymbol{v}$

$$\boldsymbol{G}^{uv}(s) = \boldsymbol{B}^T (s\boldsymbol{I} - \boldsymbol{A})^{-1} \boldsymbol{B} = \underbrace{\boldsymbol{B}_d^T (s\boldsymbol{I}_d - \boldsymbol{A}_d)^{-1} \boldsymbol{B}_d}_{=:\boldsymbol{G}_d^{uv}(s)} + \boldsymbol{G}_{nd}^{uv}(s) \tag{22}$$

is considered. The term $\boldsymbol{G}_r^{uv}$ corresponds to the neglected $n_{nd} = N - n_d$ modes. The coefficients $\mathcal{H}_k$ of a Taylor series around $s_0 = 0$

$$\boldsymbol{G}^{uv}(s) = \sum_{k=0}^{N_k} \underbrace{(-1)^{k+1} \boldsymbol{B}^T \boldsymbol{A}^{-k-1} \boldsymbol{B}}_{=:\mathcal{H}_k(0)} \, s^k = \sum_{k=0}^{N_k} \Big(\mathcal{H}_{d_k}(0) + \mathcal{H}_{nd_k}(0) \Big) s^k \tag{23}$$

are the so called 'low order moments' which are used in 'moment matching methods' for model reduction purposes [12]. Considering the l-th friction joint, the first correction term $\mathcal{H}_{r_0}$ for the second order system in (5) is given by

$$\mathcal{H}_{r_0} = \mathcal{H}_0 - \mathcal{H}_{d_0} = \boldsymbol{e}_j^T \left(\boldsymbol{K}^{-1} \boldsymbol{e}_j - \boldsymbol{\Phi}_d \boldsymbol{\Omega}_d^{-1} \boldsymbol{\Phi}_d^T \boldsymbol{e}_j \right) = \boldsymbol{e}_j^T \left(\boldsymbol{I} - \boldsymbol{\Phi}_d \boldsymbol{\Phi}_d^T \boldsymbol{M} \right) \boldsymbol{K}^{-1} \boldsymbol{e}_j \tag{24}$$

when no damping is assumed[4], $\boldsymbol{e}_l$ being the l-th column of the matrix $\boldsymbol{E}$. $\mathcal{H}_{r_0}$ represents the difference between the static deformation $\boldsymbol{K}^{-1} \boldsymbol{e}_l$ caused by an unit load acting in the l-th joint and the deformation $\boldsymbol{\Phi}_d \boldsymbol{\Omega}_d^{-1} \boldsymbol{\Phi}_d^T \boldsymbol{e}_j$ which results from the mode superposition of the modes $\boldsymbol{\Phi}_d$. By taking further coefficients $\mathcal{H}_{r_k}$ into account, the following recursive formula (Arnoldi process)

$$\boldsymbol{\psi}_k^* = \boldsymbol{K}^{-1} \boldsymbol{M} \boldsymbol{\psi}_{k-1} \tag{25}$$

$$\boldsymbol{\psi}_k^{**} = \left(\boldsymbol{I} - \boldsymbol{\Sigma}_{k-1} \boldsymbol{\Sigma}_{k-1}^T \boldsymbol{M} \right) \boldsymbol{\psi}_k^* \tag{26}$$

$$\boldsymbol{\psi}_k = \boldsymbol{\psi}_k^{**} \left(\boldsymbol{\psi}_k^{**T} \boldsymbol{M} \boldsymbol{\psi}_k^{**} \right)^{-\frac{1}{2}} \tag{27}$$

[4] Similar expressions can be given for damped second order systems [12]

for the calculation of so called Krylov vectors [13] can be derived. The starting vector is chosen as $\boldsymbol{\psi}_0 = \boldsymbol{K}^{-1}\boldsymbol{e}_j$. The matrix $\boldsymbol{\Sigma}_k = [\boldsymbol{\Phi}_d\ \boldsymbol{\psi}_1\ \dots\ \boldsymbol{\psi}_{k-1}]$ contains not only the eigenvectors of $\boldsymbol{\Phi}_d$ but the new generated vectors $\boldsymbol{\psi}_{k-1}$ $(k>1)$ as well. Thus, the bracket term in (26) and (27) constitute an orthonormalization with respect to the mass matrix. By generating K vectors for each friction joint j, a total number of $l \cdot K$ Krylov vectors is obtained. Since the columns of $\boldsymbol{\Psi} = [\boldsymbol{\psi}_1\ \dots\ \boldsymbol{\psi}_{lK}]$ are not orthogonal with respect to the stiffness matrix, a singular value decomposition (SVD)

$$\boldsymbol{\Psi}^T \boldsymbol{K} \boldsymbol{\Psi} = \boldsymbol{T}\hat{\boldsymbol{\Omega}}\boldsymbol{T} \qquad \text{where} \qquad \hat{\boldsymbol{\Omega}} = \operatorname{diag}(\hat{\omega}_m^2)\,, \quad m = 1, \dots, lK \tag{28}$$

is used to yield vectors $\hat{\boldsymbol{\Psi}} = \boldsymbol{\Psi}\boldsymbol{T}$ which also diagonalize the stiffness matrix.

When the transformation

$$\boldsymbol{z}(t) = [\boldsymbol{\Phi}_d\ \hat{\boldsymbol{\Psi}}] \begin{bmatrix} \boldsymbol{\eta}_d \\ \hat{\boldsymbol{\eta}} \end{bmatrix} \qquad \text{with} \quad \hat{\boldsymbol{\Psi}} = [\hat{\boldsymbol{\psi}}_1\ \dots\ \hat{\boldsymbol{\psi}}_{lK}] \quad \text{and} \quad \hat{\boldsymbol{\eta}} = [\hat{\eta}_1\ \dots\ \hat{\eta}_{lK}]^T\,, \tag{29}$$

is applied instead of (6), the reduced model (19) is extended by the additional rows

$$\ddot{\hat{\eta}}_m + 2\hat{\xi}_m\,\hat{\omega}_m\,\dot{\hat{\eta}}_m + \hat{\omega}_m^2\,\hat{\eta}_m = \hat{\boldsymbol{\psi}}_m^T \boldsymbol{E}\tilde{\boldsymbol{u}}_R^*(\tilde{\boldsymbol{v}}^*, \boldsymbol{\varphi}) + \hat{\boldsymbol{\psi}}_m^T \boldsymbol{F}\boldsymbol{w}\,, \qquad m = 1, \dots, lK\,. \tag{30}$$

And the expression for the approximated sliding velocity is

$$\tilde{\boldsymbol{v}}^* = \boldsymbol{E}^T \boldsymbol{\Phi}_d \dot{\boldsymbol{\eta}}_d + \boldsymbol{E}\hat{\boldsymbol{\Psi}}\dot{\hat{\boldsymbol{\eta}}}\,. \tag{31}$$

The last expression in (31) can be seen as a correction term which reduces the approximation error for low frequencies by compensating for the neglected modes of the original system. This also improves the approximated friction forces $\boldsymbol{u}_R^*$.

4.3 Example

As an example, the proposed reduction method is now applied to the 10-bay truss structure illustrated in Fig. 7. The displacements in y- and z-direction at the sensor location are used as measurements $\boldsymbol{y}$. The disturbance $\boldsymbol{w}$ is a triangular pulse in y- and z-direction which acts as a force at the mast tip as shown in Fig. 7.

For the present example a configuration with three friction joints (two of type A and one of type B) is used. Figure 8 shows the first $n = 600$ eigenvalues w_i^c for the rotational joint (type B) and for one of the translational joints (type A). As can be seen, type A mainly affects the lowest modes whereas type B also excites higher modes. Therefore, only the friction joint of type B (actuator 1) will be considered as 'active', in order to point out the effect of the correction term $\boldsymbol{E}\hat{\boldsymbol{\Psi}}\dot{\hat{\boldsymbol{\eta}}}$ in (31) on the approximation of the sliding angular velocity $\tilde{v}_1$ and the friction moment $\tilde{u}_{R_1}$. In the other two joints of type A

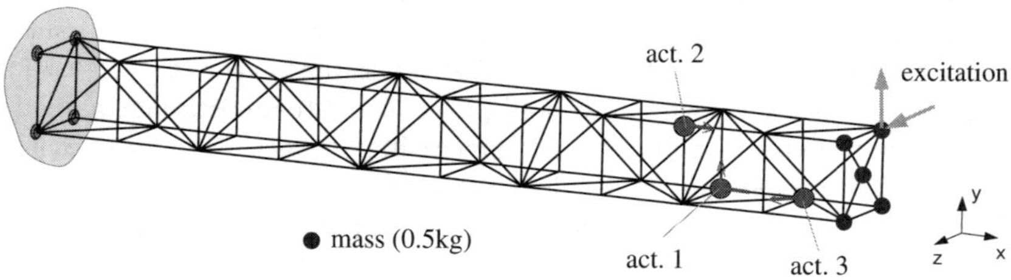

Fig. 7. Example: Structure with semi-active joints

no friction forces are transmitted and they are regarded as frictionless joint connections with zero normal force ($F_{N_{2,3}} = 0$). The reason for using those 'inactive' joints anyway is the resulting local stiffness reduction which is necessary for a significant effect of the rotational joint to occur.

Using the procedure from Sect. 4.1 $n_d = 60$ dominant modes are selected from the first $n = 100$. All higher modes are not included in the reduced model despite their large eigenvalues[5]. For a constant normal force $F_{N_1} = 400\,\mathrm{N}$, the friction moment $\tilde{u}_{R_1}$ of the reduced system given by (19) is plotted in Fig. 9 together with the friction moment obtained from the original system (5). It is obvious that the approximation of the actual friction moment is quite bad. However, augmenting the modal subspace spanned by the eigenvectors $\boldsymbol{\phi}_{d_i}$ with just two Krylov vectors $\hat{\boldsymbol{\psi}}_{1,2}$ $(l = 1\,,\; k = 1, 2)$ improves the low frequency approximation remarkably, as can be seen from the time signal of $\tilde{u}^*_{R_1}$ in Fig. 9.

5 Semi-Active Control

In the following, two different control approaches are proposed. In the first approach each semi-active friction joint has its own local feedback controller. The second concept uses one global clipped-optimal controller.

5.1 Local Feedback Control

Several semi-active control laws for friction dampers have been derived [14]. A bang-bang controller can be designed by inspection of the time derivative of the Lyapunov function representing the system energy. For the LuGre friction model defined by (2) and (1), one arrives at the feedback law

$$F_{N_j} = \begin{cases} F_{N_j}^{\max} & , \quad \mu_j \, v_j \geq \varepsilon_j \\ F_{N_j}^{\min} + (F_{N_j}^{\max}/\varepsilon_j)\, \mu_j \, v_j & , \quad 0 < \mu_j \, v_j < \varepsilon_j \\ F_{N_j}^{\min} & , \quad \mu_j \, v_j \leq 0 \end{cases} \tag{32}$$

[5] The purpose of choosing the number of modes n quite small is to point out the effect of the correction term. Usually, at least all modes lying in the frequency range of the disturbances should be considered.

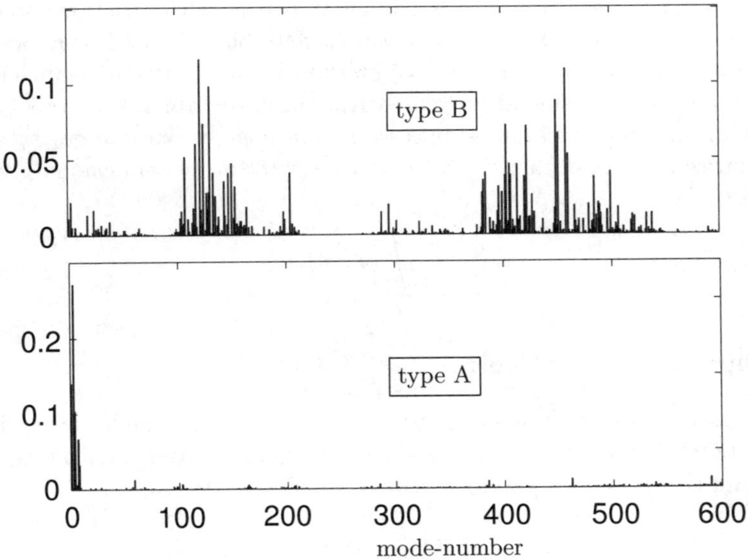

Fig. 8. Eigenvalues $w^c_{u_i}$

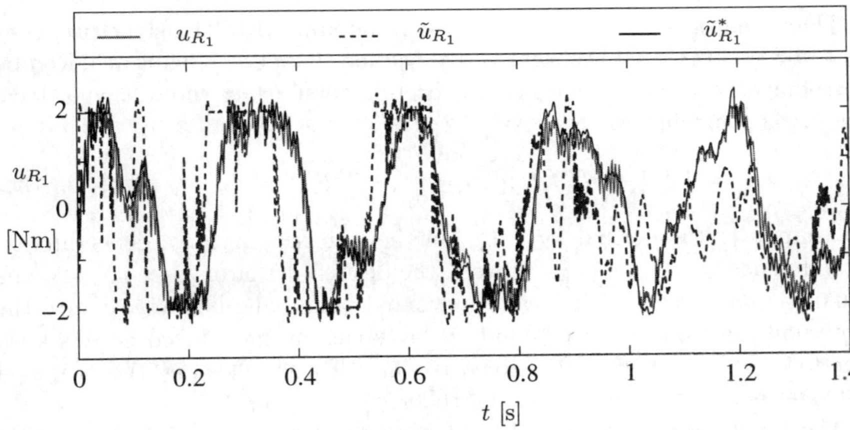

Fig. 9. Approximation of the friction moment u_{R_1}

which maximizes the energy dissipation in the adaptive joint by avoiding sticking. A boundary layer ε_j has been introduced to avoid chattering (quasi-sticking) which degrates the energy dissipation. The dynamic friction coefficient μ_j is obtained from the relation

$$\mu_j(t) = \frac{u_{R_j}(t - \Delta t)}{F_{N_j}(t - \Delta t)} \tag{33}$$

where $1/\Delta t$ is the sampling rate of the controller.

It is assumed that besides the sliding velocities $\boldsymbol{v}$ the friction forces $\boldsymbol{u}_R$ can be measured, as well. Otherwise they have to be estimated using a simple integrator model as proposed by [15] or the dynamic friction coefficients μ_j have to be observed using an extended Kalman filter [16, 17].

For an efficient vibration suppression, the control parameters ε_j of the SISO-controller are optimized [18] with respect to the system energy given by

$$E(t_e) = \int_0^{t_e} \left(\boldsymbol{x}^T \boldsymbol{x}\right) \, \mathrm{d}t \, . \tag{34}$$

5.2 Clipped-Optimal Control

For the second approach a so called clipped-optimal controller is used [19]. First, a LQG controller [20] is designed assuming an active control system and neglecting the actuator dynamics. The optimal actuator forces

$$\boldsymbol{u}_d = -\boldsymbol{K}\,\hat{\boldsymbol{x}} \tag{35}$$

are defined by the gain matrix $\boldsymbol{K}$, which is obtained as a solution of the algebraic Riccati equation [20] and the state vector $\hat{\boldsymbol{x}}$ is estimated by means of a Kalman filter [21].

The actuator forces $\boldsymbol{u}_d$ cannot directly be applied to the structure, they has to be generated by the semi-active friction dampers. This is achieved by controlling the normal force in each friction joint using the following local bang-bang controller

$$F_{N_j} = F_{N_j}^{\min} + F_{N_j}^{\max} H\left[e_j \, u_{R_j}\right] \quad , \qquad F_{N_j}^{\min} \leq F_{N_j} \leq F_{N_j}^{\max} \tag{36}$$

which updates the normal force F_{N_j} in the j-th friction force depending on the difference $e_j = u_{d_j} - u_{R_j}$ between the desired actuator force u_{d_j} and the actual friction force u_{R_j}. To avoid unwanted bang-bang behavior due to the Heavyside function $H(\cdot)$, a boundary layer can be introduced as has been done for the SISO controller defined by (32). In the following, the proposed controller is refered to as cLQG controller.

The approximation of the desired actuator forces is limited because of the dissipativity constraint and the boundedness of the friction force. If the measurement of the actual friction forces turns out to be difficult, the Kalman estimator can be extended by an integrator model [15] to estimate the friction forces, as well.

6 Simulation Results

The simulation results in this section will show the potential of the proposed semi-active approach to vibration suppression. The boundary conditions and disturbance (impulse force with an amplitude of 200 N over a time period of 40 ms) are the same as described in the previous Sect. 4.3. Three semi-active friction joints are placed at optimal locations. The optimal locations for a specific mode are determined by considering the strain energies in the end connectors which occur when the structure is deformed according to the mode shape. Since the focus has been put on the lower modes, all semi-active joints are of type A. The replaced truss nodes and the corresponding members which are adaptively connected to them are marked in Fig. 10.

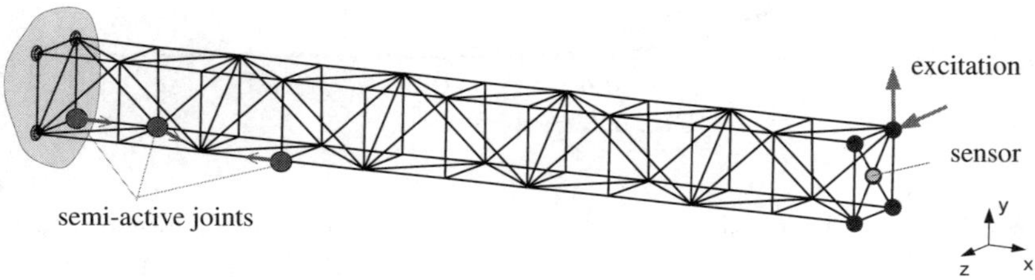

Fig. 10. Structure with three semi-active joints at optimal positions

The numerical model is reduced to 24 states using the proposed reduction method. The 24 states represent 9 modes and 3 additional 'correction modes' (Krylov vectors) generated for each friction joint.

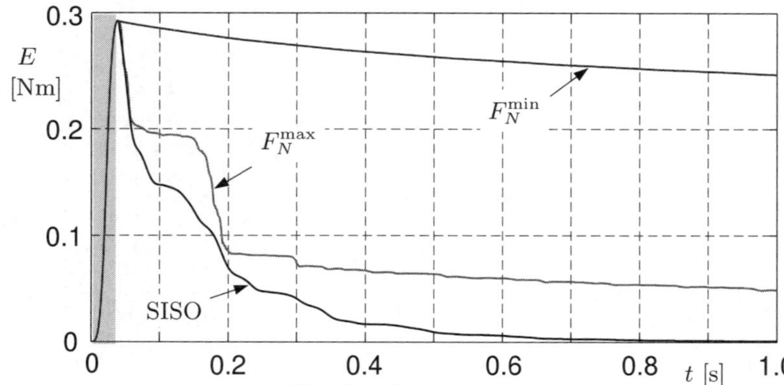

Fig. 11. System energy

Figure 11 compares the decay of the system energy E (defined by (34)) for the structure with passive and semi-active friction joints. A passive joint has a constant normal force which is set to either $F_N^{\min}$ or $F_N^{\max}$. For the semi-active joints the control law (32) with optimized control parameter ε_j is used.

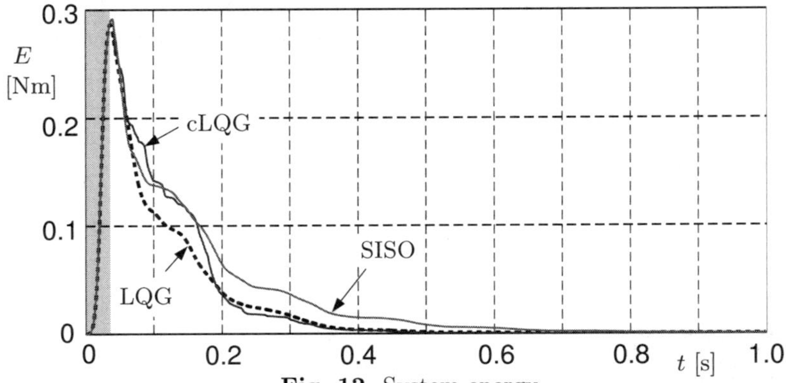

Fig. 12. System energy

It should be noted that during the excitation time period, the normal forces are set to $F_N^{\max}$ in order to enable a comparison of the different concepts. As can be seen, the damping can significantly be enhanced by controlling the normal forces in the friction joints. With a constant maximal normal force good vibration suppression is achieved at the beginning. However, as the vibration amplitude decreases sticking occurs and no energy is dissipated. Only the small structural damping reduces the vibrations.

The performance of the cLQG controller is shown in Fig. 12 together with the SISO controller. Furthermore, the response of a fully active system where the optimal actuator forces $\boldsymbol{u}_d$ are provided by the LQG controller (35) are directly applied to the structure. The corresponding tip deflections in y- and z-direction are given in Figs. 13 and 14 respectivly.

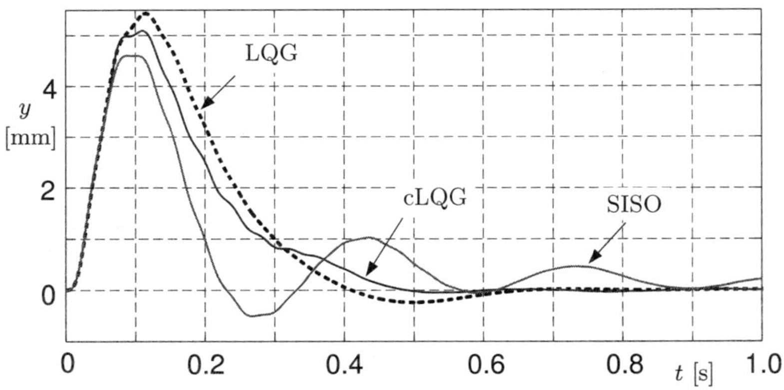

Fig. 13. Deflection of the mast tip in y-direction

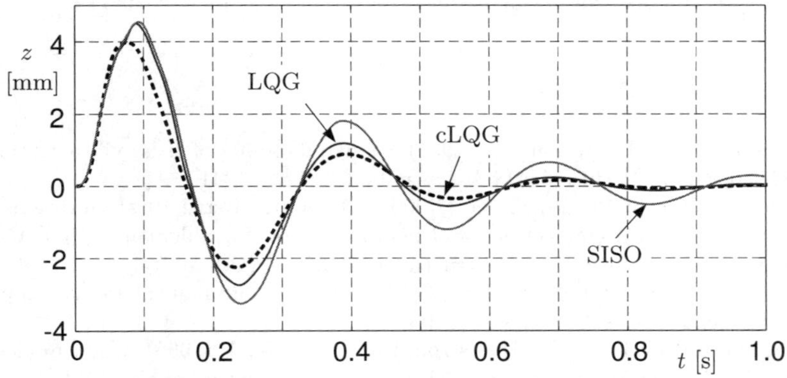

Fig. 14. Deflection of the mast tip in z-direction

When considering the system energy, the cLQG controller performs only slightly better than the three SISO controllers. The advantage of the cLQG controller becomes more obvious when the tip deflections are compared as shown in Figs. 13 and 14. In order to see how a fully active system would perform, the corresponding plots are included in Figs. 12, 13 and 14 as well. As can be seen, the suppression of the tip deflections provided by the LQG controller (active system) is only significantly better when compared to the cLQG controller (semi-active system). However, it is important to point out that the semi-active approach requires only a fraction of the control power of the active system.

7 Conclusions

The present semi-active approach to vibration suppression of large space truss structures is based on controlled energy dissipation in friction joints. A numerical model of the adaptive structure is presented, including the nonlinear dynamics of the friction joints. The finite element model of the truss structure is updated by using data from an Experimental Modal Analysis. And the parameters of the friction model are identified from measurements on an isolated friction joint. To obtain a low order model a reduction method is proposed which is a combination of balanced reduction and matching moments method. A numerical example illustrates the approximation quality of the reduced model.

Two different control approaches for semi-active damping using friction joints are introduced. Finally, the simulation results of the truss structure with three semi-active friction joints demonstrate the efficiency of the present vibration suppression approach.

Acknowledgement. The authors gratefully acknowledge the support of the research by the Deutsche Forschungsgemeinschaft (DFG) under grant GA 209/24.

References

1. Onoda J, Sano T, Minesugi K (1995) Passive damping of truss vibration using preloaded joint backlash. AIAA Journal 33(7):1335–1341
2. Gaul L, Albrecht H, Wirnitzer J (2001) Damping of structural vibrations using adaptive joint connections and neural control, In: Suleman A, (ed) CISM Courses and Lectures: Smart Structures 429:86–97 (Udine, Italy)
3. Gaul L, Nitsche R (2001) The role of friction in mechanical joints. Applied Mechanics Reviews (ASME) 54(2):93–106
4. Canudas de Wit C, Olsson H, Åström KJ, Lischinsky P (1995) A new model for control of systems with friction, IEEE Trans. Automatic Control 40(3):419–425
5. Armstrong-Hélouvry B, Dupont P, Wit CD (1994) A survey of models, analysis tools and compensation methods for the control of machines with friction. Automatica 30(7):1083–1138
6. Hanss M, Oexl S, Gaul L (2002) Simulation and analysis of structural joint models with uncertainties. In: Proc. of the International Conference on Structural Dynamics Modeling – Test, Analysis, Correlation and Validation. pp. 165–174
7. Kukreti AR, Issa HI (1984) Dynamic analysis of nonlinear structures by pseudonormal mode superposition method. Computers & Structures 19(4):643–663
8. Friswell MI, Mottershead JE (1995) Finite Element Model Updating in Structural Dynamics. Kluwer Academic Publishers, Dordrecht
9. Moore BC (1981) Principal component analysis in linear systems: Controllability, observability, and model reduction. IEEE Trans. Automatic Control vol. AC-26(1):17–32
10. Gawronski WK, Juang J-N (1990) Model reduction for flexible structures. Control and Dynamic Systems: Advances in Theory and Application 36:143–222
11. Gregory Jr. CZ (1984) Reduction of large flexible spacecraft models using internal balancing theory. J. Guidance and Control 7(6):725–732
12. Craig Jr. RR, Su T-J (1991) Model reduction and control of flexible structures using krylov vectors. J Guidance and Control 14(2):260–267
13. Bai Z (2002) Krylov subspace techniques for reduced-order modeling of large-scale dynamical systems. J Appl. Num. Math. 43:9–44
14. Lane JS, Ferri AA, Heck BS (1992) Vibration control using semi-active friction damping. In: Friction-induced vibration, chatter, squeal, and chaos. ASME 49:165–171
15. Hu R, Müller PC (1997) Position control of robots by nonlinearity estimation and compensation: Theory and experiements. J. Intelligent and Robotic Systems 20:195–209
16. Gaul L, Nitsche R (2000) Friction control for vibration suppression. Mechanical Systems and Signal Processing 14(2):139–150
17. Gaul L, Nitsche R (2003) Lyapunov design of damping controllers. Arch. Appl. Mech. 72:865–874
18. Wirnitzer J, Kistner A, Gaul L (2002) Optimal placement of semi-active joints in large space truss structures. In: Agnes GS (ed) SPIE Conference on Smart Structures and Materials 2002: Damping and Isolation No.4697:246–257, (San Diego, USA)

19. Dyke SJ et.al. (1996) Lyapunov design of damping controllers. Arch. Appl. Mech. 5:565–575
20. Kwakernaak H, Sivan R (1972) Linear Optimal Control Systems, Wiley, New York
21. Kalman RE, Bucy R (1961) New results in linear filtering and prediction. ASME J. Basic Engineering 83:95–108

Design of Adaptive Structures under Random Impact Conditions

J. Holnicki-Szulc, P. Pawłowski and M. Wikło

Institute of Fundamental Technological Research PAS
Świętokrzyska 21, 00–049 Warsaw, Poland
holnicki@ippt.gov.pl

Summary. The methodology (based on the so-called Dynamic Virtual Distortion Method) of the design of structures exposed to impact loading is presented in the work. Minimization of material volume and accelerations of structural response are chosen as the objective functions for optimal design of structures adapting to impact loads. The cross-sections of structural members as well as stress levels triggering plastic-like behavior and initial prestressing are the design parameters. A general formulation of this problem, as well as particular cases, are discussed.

Key words: adaptive structures, optimal control, dynamic sensitivity analysis

1 Introduction – Problem Formulation

Motivation for the undertaken research is to respond to requirements for high impact energy absorption e.g. in the structures exposed to the risk of extreme blast, light, thin wall tanks with high impact protection, vehicles with high crashworthiness, protective barriers, etc. Typically, the suggested solutions focus on the design of passive energy absorbing systems. These systems are frequently based on the aluminum and/or steel honeycomb packages characterized by a high ratio of specific energy absorption. However high is the energy absorption capacity of such elements, they still remain highly redundant structural members, which do not carry any load in the actual operation of a given structure. In addition, passive energy absorbers are designed to work effectively in pre-defined impact scenarios. For example, the frontal honeycomb cushions are very effective during a symmetric axial crash of colliding cars, but are completely useless in other types of crash loading. Consequently, distinct and sometimes completely independent systems must be developed for specific collision scenarios.

In contrast to the standard passive systems, the proposed approach focuses on active adaptation of energy absorbing structures (equipped with a sensor

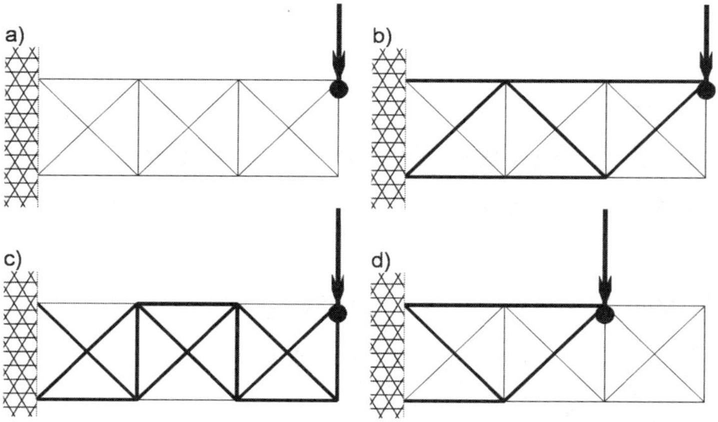

Fig. 1. Testing example of truss-like cantilever. a) the initial configuration, b) the stiffest substructure, c) the most compliable substructure, d) the stiffest substructure designed for the second load case VDM Based Dynamic Analysis of Adaptive Structure

system detecting impact in advance and controllable semi-active dissipaters, so called structural fuses) with a high ability of adaptation to extreme overloading. The concept formulation and first numerical analysis are based on the previously published paper [4]. Various formulations of crashworthiness-based structural design problem are presented in papers [1, 2, 8, 9, 10, 11–14], while the adaptive crashworthiness concept has been first proposed in [3] and [6]. The optimal design methodology proposed below combines sensitivity analysis with the redesign process, allowing optimal redistribution of material as well as stress limit control in structural fuses. It is assumed that this 'smart' devices are able to release structural connections in a controlled way, triggering plastic-like distortions mimicking elasto-plastic behaviour shown in Fig. 2.

The objectives in design of impact absorbers are the following:

- to sustain all expected impact loads and especially the one with maximal impact energy,
- to absorb all expected impact loads in the most smooth way, minimizing maximal accelerations.

Assuming first only one state of impact load, with possible variation of impact intensity, let us take into account the following definitions:

E^u – maximal expected impact energy

σ^u – yield stress level for ideal elasto-plastic material used to built the structure

β^u – maximal allowed plastic-like distortion to be generated in structural fuses.

Now, assume that in hiperstatic truss-like structure (k redundant) k independent states of self-equilibrated stresses can be generated in such a way that the yield stress level $\sigma^\star \leq \sigma^u$ is reached (under a given load) at the same time in all structural members. Then, the minimal volume of material V necessary to dissipate the maximal expected impact energy E^u can be calculated from the following formula: $E^u = V\sigma^u\beta^u$, requiring uniformly distributed maximal stresses triggering plastic distortions in all structural members. Note, that some initial state of virtual distortions $\beta^{0\prime}$ has to be introduced in order to generate desired prestress described above.

Having defined the minimum volume of structural material let us discuss the heuristic methodology for structural remodeling on the example of truss-like cantilever shown in Fig. 1a. and loaded dynamically in structural node (mass m hits the tip joint with the initial velocity v). The well-known solution for the optimal topology of the stiffest substructure S_1 (made from the determined material volume) is shown in Fig. 1b (cf. [7]) while the most compliant substructure S_2 is demonstrated in Fig. 1c. Both of these substructures are isostatic and the material distribution (sizing of members' cross-sections) guarantees uniformly distributed locally extremal stresses in response to the impact $I(m, v)$.

Let us now calibrate the substructures S_1 and S_2 defined above in such way, that the volumes of the material for both of them are identical and equal to $V/2$. Finally, let us compose the resultant structure $(S_1 \cup S_2)$, where the cross-section of each element i is defined as the following mean value: $\mu_i = (\mu_i^1 + \mu_i^2)/2$, and where the coefficient μ_i^1 ($\mu_i^1 = A_i/A_i'$, A_i and A_i' denote modified and the initial cross-section of the member i, respectively) describes the ratio of cross-section modification for the element i from the substructure S_1, while a coefficient μ_i^2 describes the ratio of cross-section modification for the element i from the substructure S_2. In the example shown in Fig. 1 the topology of the resultant complex structure is the same as the topology of the initial configuration (Fig. 1a).

The **remodeling** process described above is related to the following **first problem** formulation:

$$\min V = \min \sum_{\mathrm{i}} \mu_i A_i' l_i \tag{1}$$

subject to the following constraints:

$$\begin{gathered} \left|\beta_i^0(t)\right| \leq \beta^u \\ |\sigma_i(t)| \leq \sigma^\star \leq \sigma^u \\ \beta_i^0(t)\sigma_i(t) \geq 0 \end{gathered} \tag{2}$$

where l_i denotes the length of the member i and stresses $\sigma_i(t)$ depend on the maximal expected impact load $I(m, v)$ and the control parameters: μ_i, $\sigma_i^\star$, $\beta_i^{0\prime}$, what will be discussed in the next section. The constraint (2) describes the condition of dissipative character of plastic-like distortions generation.

Having determined the material remodeling described above, the following, **second problem of adaptation** to the identified (in advance, using a sensor system) impact load can be formulated:

$$\min \max \ddot{u}_i(t) \tag{3}$$

subject to the following constraints:

$$|u_i(t)| \leq u^u \tag{4}$$

where displacements $u_i(t)$ depend on the identified in real time impact load $I(m, v)$, previously determined material redistribution μ_i and the control parameters: $\sigma_i^\star$, $\beta_2^{0\prime}$ describing plastic-like adaptation and the initial prestress, what will be discussed in the next section.

The following strategies of adaptation to the identified in real time impact loads have to be considered:

- detachment of elements i: $(i \in S_2) \wedge (i \notin S_1)$ and tuning proper stress levels $\sigma^\star$ allows to obtain the most compliable substructure S_2, able to receive (the most smoothly) small impacts with constraint (4) still satisfied,
- detachment of elements i: $(i \in S_1) \wedge (i \notin S_2)$ and tuning proper stress levels $\sigma^\star$ allows to obtain the stiffest substructure S_1, able to receive stronger impacts without excessive deflections violating constraint (4),
- introducing proper initial distortions $\beta_2^{0\prime}$ in elements of the substructure S_1, and tuning stress levels $\sigma^\star$ allows to obtain the uniform distribution of the stress level $\sigma^\star$ in all structural elements and simultaneous triggering plastic-like distortions, what leads to the most effective impact energy dissipation. This third strategy should be applied when the impact capacity of the second one is too low.

Note, that *detachment* of structural elements mentioned in the two first strategies can be realized introducing properly determined initial distortions $\beta_2^{0\prime}$.

In multi-load cases compromise solutions for the material redistribution have to be taken into account. For example, the same impact $I(m, v)$ provoked by different masses and initial velocities leads to different mass distribution (μ_i). Optimal solutions for slow impacts (bigger mass and lower initial velocities) require mass transportation to the support neighborhood while in the case of fast impact (the same impact energy provoked by smaller mass hitting the structure with higher velocity) the mass transportation to the neighborhood of loaded node can be observed (cf. [7]). Therefore, in case of expected variable m/v ratio, the optimal material redistribution requires a compromise solution. Considering optional impact load state (cf. Fig. 1d), the topology of the stiffest substructure is different from the previous one (Fig. 1b), what also demonstrates that the determinations of the compromise solution for a multi-load remodeling problem require new numerical tools able to tackle this challenging problem. The proposition of new numerical technique (based on so called Virtual Distortion Method, VDM, cf. [5]) able to solve the problem discussed above will be discussed in the next two sections.

2 VDM Based Dynamic Analysis of Adaptive Structure

In this chapter we will formulate the VDM based description of the dynamic response of elasto-plastic truss structure. Applying discretized time description, the evolution of strains and stresses (with respect to initial cross-sections) can be expressed as follows:

$$\varepsilon_i(t) = \varepsilon_i^L(t) + \sum_{\tau \le t} \sum_j D_{ij}(t-\tau) \cdot \varepsilon_j^0(\tau) + \sum_{\tau \le t} \sum_k D_{ik}(t-\tau) \cdot \beta_k^0(\tau) \quad (5)$$

$$\begin{aligned} \sigma_i'(t) =& E_i \left(\varepsilon_i(t) - \varepsilon_i^0(t) - \beta_i^0(t) \right) \\ \sigma_i'(t) =& E_i \Bigg[\varepsilon_i^L(t) + \sum_{\tau \le t} \sum_j D_{ij}(t-\tau) \cdot \varepsilon_j^0(\tau) - \varepsilon_i^0(t) + \\ & \sum_{\tau \le t} \sum_k D_{ik}(t-\tau) \cdot \beta_k^0(\tau) - \beta_i^0(t) \Bigg] \end{aligned} \quad (6)$$

where so called dynamic influence matrices $D_{ij}(t-\tau)$ describe the strain evolution caused in the truss element member i and in time instance t, due to unit virtual distortion impulse generated in member j in the time instant τ. The vector $\varepsilon_i^L(t)$ denotes the strain evolution due to external loads applied to the elastic structure with initial material distribution (unmodified cross-sections of members), $\varepsilon_i^0(t)$ denotes virtual distortions responsible for modification of design variables and $\beta_i^0(t)$ describes plastic-like distortions. Note that the matrix D stores information about the properties of the entire structure (including boundary conditions) and describes dynamic (not static) structural response to a locally generated impulse of virtual distortion. Note also that the influence of local modifications of design variables on the stiffness matrix only, was assumed in further analysis. The full analysis taking into account the

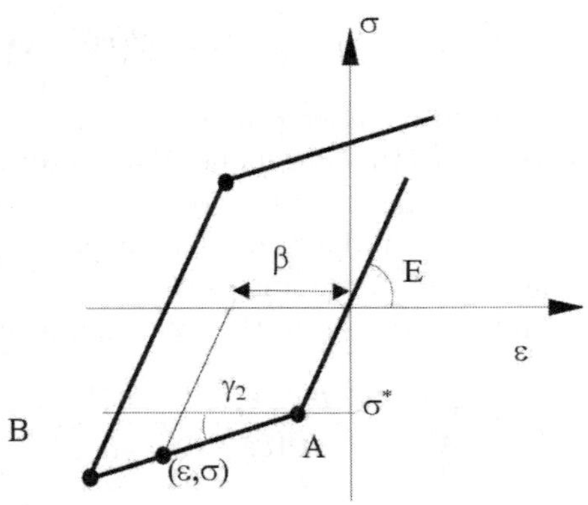

Fig. 2. Piece-wise linear constitutive relation for the adaptive structural member

influence of virtual distortions $\varepsilon_i^0(t)$ on both, the stiffness as well as the mass matrices is more complicated and will be discussed in separate publication. From now on, we assume that small Latin index j runs through all modified members, and small Latin index k runs through all plastified elements.

In order to take into account elasto-plastic structural behaviour, let us use the bilinear constitutive model with hardening (Fig. 2), given by the equation (7)

$$\sigma_i(t) - \sigma_i^\star = \gamma_i E_i \left(\varepsilon_i(t) - \varepsilon_i^\star\right) \tag{7}$$

where $\sigma_i^\star$ denotes plastic yield stress, γ_i denotes hardening parameter and E_i denotes Young's modulus.

Now, when we substitute stress (6) and strain (5) evolution in time to the formula (7) we obtain the following equation:

$$\begin{aligned}\beta_i^0(t) = (1-\gamma_i)\left(\varepsilon_i^L(t) - \varepsilon_i^\star\right) + (1-\gamma_i)\sum_{\tau\le t}\sum_j D_{ij}^D(t-\tau)\cdot\varepsilon_j^0(\tau)\\ + (1-\gamma_i)\sum_{\tau\le t}\sum_k D_{ik}^H(t-\tau)\cdot\beta_k^0(\tau)\end{aligned} \tag{8}$$

Taking advantage of two expressions for the internal forces applied to the so called distorted (9) (with modification of material distribution modeled through virtual distortions) and modified (10) (with redesigned cross-sections from A to A', without imposing virtual distortions) structure:

$$P_i(t) = E_i A_i \left(\varepsilon_i(t) - \varepsilon_i^0(t) - \beta_i^0(t)\right) \tag{9}$$

$$P_i(t) = E_i A_i' \left(\varepsilon_i(t) - \beta_i^0(t)\right) \tag{10}$$

A formula combining components $\varepsilon_i^0(t)$ and $\beta_i^0(t)$ can be derived, where these components are non zero only for distorted and/or plastified elements.

If we assume that forces and strains in both structures: *distorted* (9) and *modified* (10) are the same, the modifications simulated with virtual distortion can be combined with these distortions through the flowing formula:

$$\varepsilon_i^0(t) = (1-\mu_i)\left(\varepsilon_i(t) - \beta_i^0(t)\right) \tag{11}$$

where $\varepsilon_i(t)$ describes strain in member i in time t, while $\mu_i = A_i/A_i'$ denotes ratio of the new cross-section to the initial one. Parameter $\mu_i \in \langle 0, 1\rangle$ specifies the size of modification of cross-sections in element i. If $\mu_i = 1$ that means that in element i the cross-section does not change, and if $\mu_i = 0$ that means that element i can be neglected in the analysis.

The formula (11) can be rewritten in the following form (12):

$$\mu_i = \frac{A_i}{A_i'} = \frac{\varepsilon_i(t) - \varepsilon_i^0(t) - \beta_i^0(t)}{\varepsilon_i(t) - \beta_i^0(t)} \tag{12}$$

Now let us substitute strain evolution in time (5) to formula (12) getting the following set of equations:

$$\varepsilon_i^0(t) = (1-\mu_i)\,\varepsilon_i^L(t) + \sum_{\tau\le t}\sum_j D_{ij}^D\,(t-\tau)\cdot\varepsilon_j^0\,(\tau)$$
$$+\sum_{\tau\le t}\sum_k D_{ik}^H\,(t-\tau)\cdot\beta_k^0\,(\tau) - \beta_i^0(t) \quad (13)$$

Note that the equations (8) are not dependant on the virtual distortions responsible for modification of design variables in time t, but only on the distortions in previous time steps $\varepsilon\,(\tau)$ because of the assumption (14).

$$D_{ij}\,(0) = 0 \quad (14)$$

Therefore, the plastic-like distortions $\beta_i^0(t)$ should be calculated first in each time step of the algorithm.

Equations (8) and (13) need only computation of the right-hand side expressions, and we need not solve the coupled sets of equations.

Formulas (8) and (13) allow us to compute the virtual distortions' development in time, modeling both: assumed remodeling of material distribution as well as adapted plastic-like stress limits.

If there is no plasticity in our problem, then plastic-like distortions are equal to zero and the equation (13) takes the following form:

$$\varepsilon_i^0(t) = (1-\mu_i)\,\varepsilon_i^L(t) + \sum_{\tau\le t}\sum_j D_{ij}\,(t-\tau)\cdot\varepsilon_j^0\,(\tau). \quad (15)$$

Analogously, if there is no remodeling, distortions are equal to zero (the parameter μ_i is equal to one) and equation (8), determining plastic like distortions' development takes the following form:

$$\beta_i^0(t) = (1-\gamma_i)\left(\varepsilon_i^L(t) - \varepsilon_i^\star\right) + (1-\gamma_i)\sum_{\tau\le t}\sum_k D_{ik}^H\,(t-\tau)\cdot\beta_k^0\,(\tau) \quad (16)$$

To prove that the VDM method gives the same solutions as commercial programs let us compare results for a simple truss structure shown in Fig. 3 with the structural response determined with ANSYS.

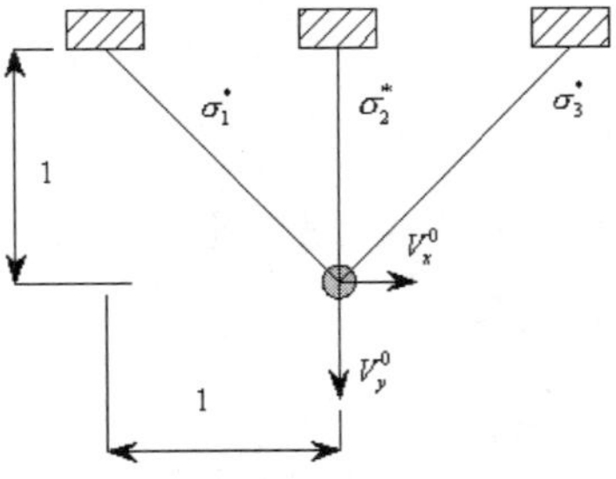

Fig. 3. Testing example truss structure. (Young modulus $E_i = 2.1\times10^11\,\mathrm{Pa}$, cross-sections $A_i = 1\times10^{-4}$ m, density $\rho_i = 7800\,\mathrm{kg/m^3}$)

All elements have different yield stress limits $\sigma_i^\star$ as well as parameters responsible for modification of design variables μ_i.

$$\sigma_1^\star = 8 \times 10^7 \text{Pa}, \qquad \sigma_2^\star = 4 \times 10^7 \text{Pa}, \qquad \sigma_3^\star = 6 \times 10^7 \text{Pa}$$
$$\mu_2 = 0.7, \qquad \mu_1 = 0.5, \qquad \mu_3 = 0.9$$

In the lower node, concentrated mass 20 kg is added, together with the following initial condition (modeling with external object):

$$V_x^0 = 3\,\text{m/s}, \quad V_y^0 = 5\,\text{m/s}.$$

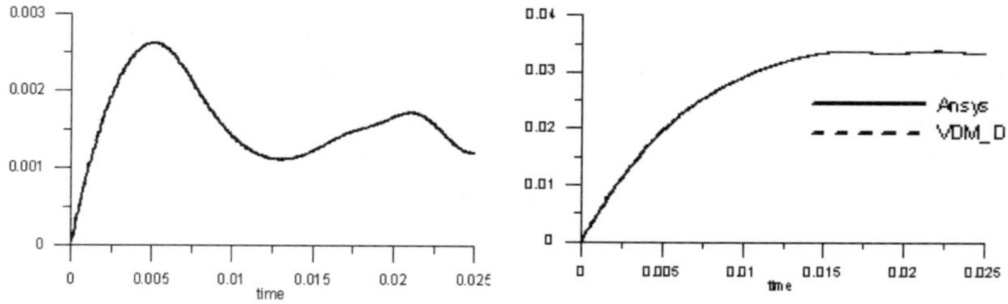

Fig. 4. Strain evolution in time for elements: a) left element, b) central element

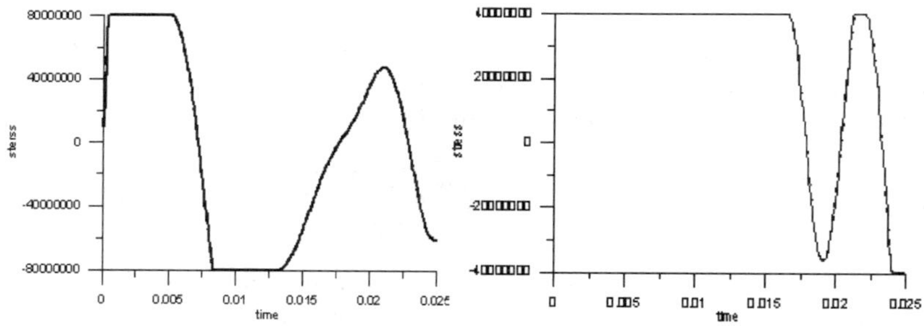

Fig. 5. Stress evolution in time for elements: a) left element, b) central element

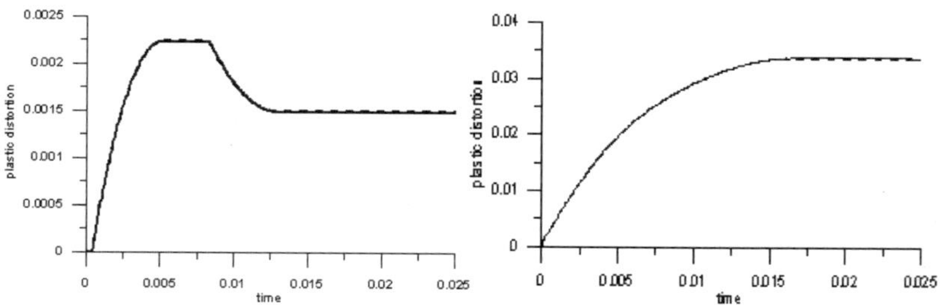

Fig. 6. Plastic distortion evolution in time for elements: a) left element, b) central element

On the graphs, the comparison of strain (Fig. 4), stress (Fig. 5) and plastic distortion (Fig. 6) development for the first and the second element, respectively is demonstrated.

Note, that only modification of the stiffness matrix (due to remodeling) has been taken into account in the above formulas. Analogous modification of the mass matrix has to be added in order to describe the complex remodeling phenomena. However, it was decided to keep this presentation simpler without additional complication of formulas.

3 Gradient Based Approach to Structural Remodeling

Let us demonstrate now, how the VDM based approach supports sensitivity analysis, useful in optimization procedures. Assume that the objective function is defined as maximization of dissipated energy during the adaptation process, given by the following formula:

$$U \max = \sum_t \sum_i \sigma_i(t) \Delta\beta_i(t) \mu_i A_i' l_i \tag{17}$$

subject to constrains:

$$-\tilde{\beta}_i \leq \beta_i \leq \tilde{\beta}_i, \quad 0 \leq \mu_i \leq 1, \quad \mu_i A_i' l_i = \text{const} \tag{18}$$

where $\tilde{\beta}_i$ denotes lower and upper limit imposed on plastic-like distortions, $\mu_i A_i' l_i$ denotes total volume of material which should be constant during the remodeling process. In this way, solving the above problem (17) for increasing load intensities, one can find the desired design of the structure with maximal impact load capacity, constraining considerations to small deformations.

To make further analysis more communicative, let us distinguish two particular cases. The first one, establishing the best material redistribution in all structural members leads to the determination of the following design variables: $\mu_i = A_i / A_i'$. In this case $\Delta\beta_k^0(t) = 0$, and $\Delta\beta_k^0(t)$ is replaced by $\varepsilon_i(t)$ in the objective function (17).

The gradient of this objective function can be calculated analytically and takes the following form:

$$\frac{\partial U}{\partial \mu_m} = \left(\frac{\partial U}{\partial \sigma_i(t)} \frac{\partial \sigma_i(t)}{\partial \varepsilon_j^0(t)} + \frac{\partial U}{\partial \varepsilon_i(t)} \frac{\partial \varepsilon_i(t)}{\partial \varepsilon_j^0(t)} \right) \frac{\partial \varepsilon_j^0(t)}{\partial \mu_m} + \frac{\partial U}{\partial \mu_m} \tag{19}$$

where the particular components can be expressed as follows:

$$\frac{\partial U}{\partial \sigma_i(t)} = \sum_i \Delta\beta_i(t)\mu_i A_i' l_i$$

$$\frac{\partial U}{\partial \varepsilon_i(t)} = \sum_i \sigma_i(t)\mu_i A_i' l_i$$

$$\frac{\partial U}{\partial \mu_m(t)} = \sum_t \sigma_m(t)\varepsilon_m(t) A_m' l_m \tag{20}$$

$$\frac{\partial \sigma_i(t)}{\partial \varepsilon_j^0(t)} = - E_i \delta_{ij}$$

$$\frac{\partial \varepsilon_j^0(t)}{\partial \mu_m} = - \varepsilon_i^L(t) + \sum_{\tau \le t} \sum_j D_{ij}^D (t-\tau) \cdot \frac{\partial \varepsilon_j^0(\tau)}{\partial \mu_m}$$

In the second case we are looking for optimal distribution of yield stress limits in structural members and the design variables during optimization process are $\sigma_i^\star$. The corresponding gradient of the objective function, with respect to yield stress limits, takes the following form:

$$\frac{\mathrm{d}U}{\mathrm{d}\sigma_l^\star} = \left(\frac{\partial U}{\partial \sigma_p(t)} \frac{\partial \sigma_p(t)}{\partial \Delta\beta_k^0(t)} + \frac{\partial U}{\partial \Delta\beta_k^0(t)} \right) \frac{\partial \Delta\beta_k^0(t)}{\partial \sigma_l^\star} \tag{21}$$

where the new components of gradient can be expressed as follows:

$$\frac{\partial \sigma_p(t)}{\partial \Delta\beta_k^0(t)} = -E_p \Delta\beta_p^0(t)\delta_{pk} \tag{22}$$

$$\frac{\partial \Delta\beta_k^0(t)}{\partial \sigma_l^\star} = -\frac{(1-\gamma_i)}{E} + (1-\gamma_i) \sum_{\tau \le t} \sum_k D_{ik}^H (t-\tau) \cdot \frac{\partial \beta_k^0(\tau)}{\partial \sigma_l^\star} \tag{23}$$

Finally, the last case couples optimization sub-problems: *remodeling* end *adaptation* of adaptive structure. The design variables describe simultaneously material redistribution as well as yield stress limits: $\mu_i = A_i/A_i'$ and $\sigma_i^\star$, respectively.

The coupled gradient formula takes the following form:

$$\frac{\mathrm{d}U}{\mathrm{d}\sigma_l^\star} = \left(\frac{\partial U}{\partial \sigma_p(t)} \left(\frac{\partial \sigma_p(t)}{\partial \varepsilon_j^0(t)} \frac{\partial \varepsilon_j^0(t)}{\partial \Delta\beta_k^0(t)} + \frac{\partial \sigma_p(t)}{\partial \Delta\beta_k^0(t)} \right) \right.$$
$$\left. + \frac{\partial U}{\partial \Delta\beta_k^0(t)} + \frac{\partial U}{\partial \mu_m} \frac{\partial \mu_m}{\partial \Delta\beta_k^0(t)} \right) \frac{\partial \Delta\beta_k^0(t)}{\partial \sigma_l^\star} \tag{24}$$

4 Design of Adaptive Structure (Small Deformations) – Testing Example

The coupled, gradient based remodeling problem described above is restricted to small deformations. However, it can be used as numerical tool for redesign of topology of adaptive structure on the base of analysis performed in the initial phase of structural response to the impact.

Let us demonstrate effectiveness of the proposed design concept of adaptive impact absorber on the base of simple truss structure shown in Fig. 3. The described above VDM technique (in development) will be soon applied to solve real design problems for the proposed smart impact absorbing type of structures.

Let us assume, that the testing structure of shock absorber (Fig. 3) is designed to absorb the impact of mass $m = 745\,\mathrm{kg}$ hitting (and sticking to the node afterwards) the node with its initial velocity v, where $v < 8.7\,\mathrm{m/s}$. It means that the maximal expected impact energy can reach $E^u = mv^2/2 = 28194.525$ J. Consequently, the material volume used to built the structure should be calculated from the formula: $E^u = V\sigma^u\beta^u$ (cf. section 1) and is equal, assuming $\sigma^u = 5 \times 10^8\,\mathrm{Pa}$ and $\beta^u = 0.12$, to $V = 5.22 \times 10^{-5}\,\mathrm{m}^2$. Now, following the methodology proposed in section 1, let us decompose the structure into two substructures: S_1 and S_2 with the same volume of material: $V_1 = V_2 = V/2$. The structural dynamic responses for substructures S_1, S_2 and for the composed structure $S_1 \cup S_2$ are shown in Figs. 7b and 7c, respectively. Changing initial impact velocity in the range: $\langle 0.54\,\mathrm{m/s}, 8.7\,\mathrm{m/s}\rangle$ we can observe evolution of stresses, strains and plastic distortions in elements of the corresponding substructures S_1 and S_2 (Fig. 7b) and also the analogous evolutions in the compound structure $S_1 \cup S_2$ prestressed by introducing initial, plastic-like distortion $\beta_2^{0\prime} = 2.62 \times 10^{-3}$ into the vertical member of the structure (Fig. 7c).

a)

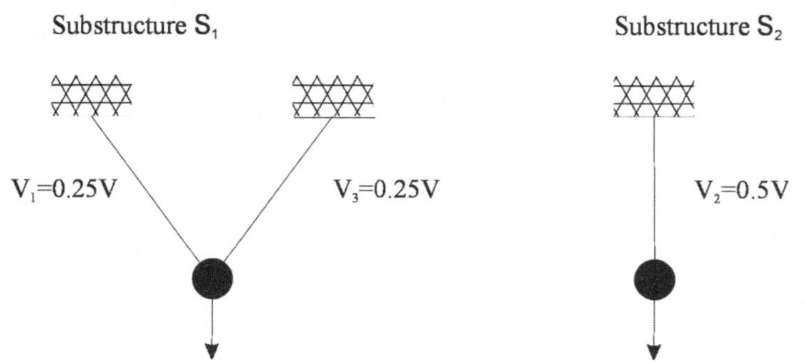

b)

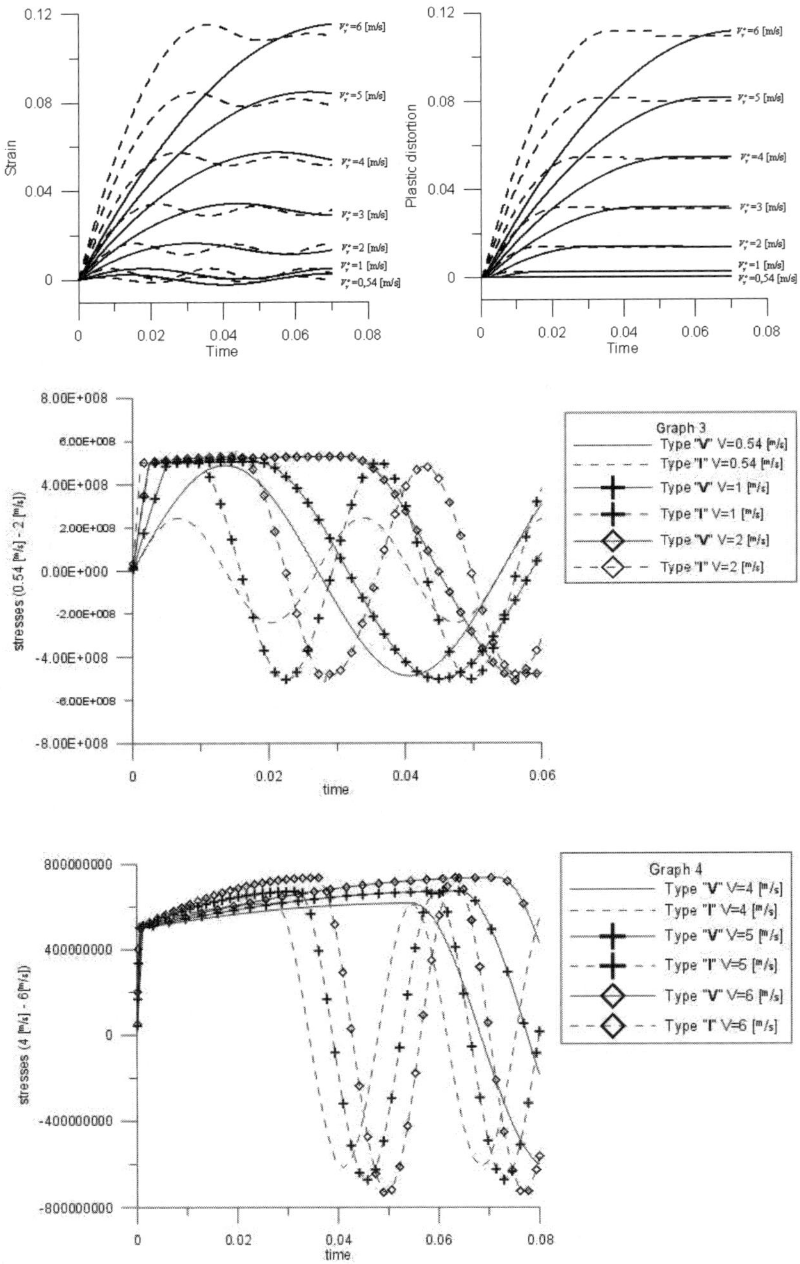

Strain
Time
Plastic distortion
Time
Graph 3
Type "V" V=0.54 [m/s]
Type "I" V=0.54 [m/s]
Type "V" V=1 [m/s]
Type "I" V=1 [m/s]
Type "V" V=2 [m/s]
Type "I" V=2 [m/s]
stresses (0.54 [m/s] - 2 [m/s])
time
Graph 4
Type "V" V=4 [m/s]
Type "I" V=4 [m/s]
Type "V" V=5 [m/s]
Type "I" V=5 [m/s]
Type "V" V=6 [m/s]
Type "I" V=6 [m/s]
stresses (4 [m/s] - 6[m/s])
time

c)

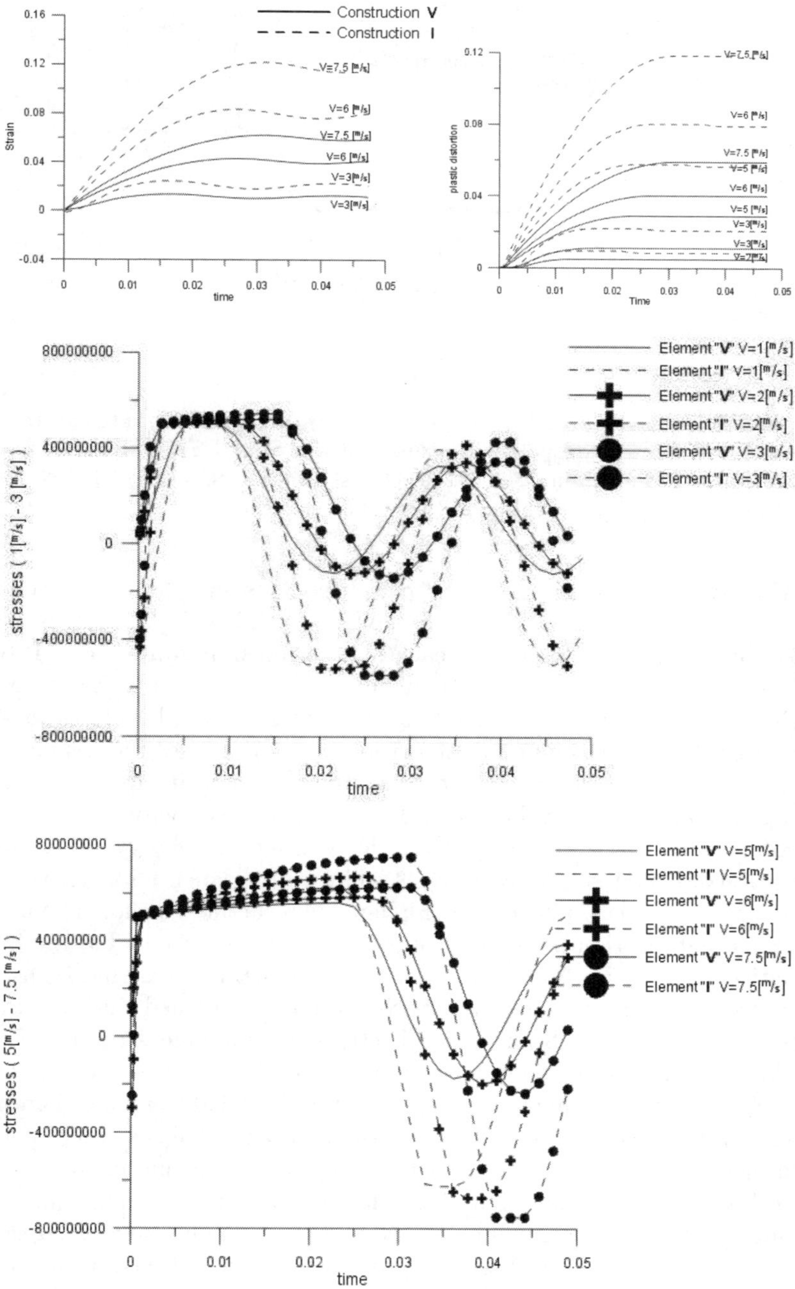
Construction V
Construction I
Strain
time
plastic distortion
Time
Element "V" V=1[m/s]
Element "I" V=1[m/s]
Element "V" V=2[m/s]
Element "I" V=2[m/s]
Element "V" V=3[m/s]
Element "I" V=3[m/s]
stresses (1[m/s] - 3 [m/s])
time
Element "V" V=5[m/s]
Element "I" V=5[m/s]
Element "V" V=6[m/s]
Element "I" V=6[m/s]
Element "V" V=7.5[m/s]
Element "I" V=7.5[m/s]
stresses (5[m/s] - 7.5 [m/s])
time

d)

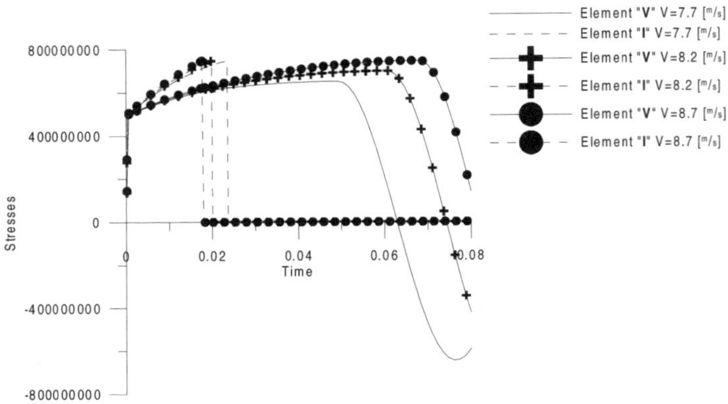

Fig. 7. Testing example of adaptive structure. a) decomposition into substructures S_1 and S_2, b) dynamic responses for substructures S_1 and S_2, c) dynamic response for the compound structure $S_1 \cup S_2$, d) stress evolution in time for the compound structure $S_1 \cup S_2$ with switching off the middle element

The following characteristic impact velocities can be identified: $v = 0.54\,\mathrm{m/s}$, when the substructure S_2 starts to yield, $v = 3.9\,\mathrm{m/s}$, when the deflection of the substructure S_2 reaches its assumed limitation $u^u = 0.108\,\mathrm{m}$, $v = 7.5\,\mathrm{m}$, when the plastic like distortion generated in substructure S_1 reaches its assumed limitation β^u, $v = 8.7\,\mathrm{m/s}$, when the plastic like distortion generated in compound structure $S_1 \cup S_2$ reaches its assumed limitation β^u. Then, analyzing numerical results presented in the Fig. 7, the following interpretation can be given. For the impact velocities $v < 0.54\,\mathrm{m/s}$ the adaptive structure responds elastically, without any energy dissipation, as the most compliant substructure S_2 is still too stiff. For the impact velocities $0.54\,\mathrm{m/s} < v < 3.9\,\mathrm{m/s}$ the most compliant substructure S_2 should be used for the impact absorption (substructure S_1 should be detached). For the impact velocities $3\,\mathrm{m/s} < v < 6.1\,\mathrm{m/s}$ the stiffest substructure S_1 should be used for the impact absorption (substructure S_2 should be detached). For the impact velocities $6\,\mathrm{m/s} < v < 8.7\,\mathrm{m/s}$ the prestressed compound structure $S_1 \cup S_2$ should be used for the impact absorption. Note, that the first mode of operation (S_2 substructure) allows to reach accelerations 280% smaller than in the second mode case (S_1 substructure) for small impacts. On the other hand, the third mode of operation ($S_1 \cup S_2$ structure) allows reaching impact capacity (in terms of energy dissipation) 110% higher than in the second mode case (S_1 substructure) for strong impacts. Note, that the limitation for impact absorption in the last case is due to active constraint β^u reached in the vertical member (cf.Fig. 7c). Then, switching off the connection between this member and the rest of the structure (cf.Fig. 7d) (semi-active adaptation in real time) allows extension of the structural impact capacity up to above 203% (comparing with S_1 substructure capacity) and the corresponding impact velocity, as the inclined members will still continue process of dissipation.

The crucial points in the proposed methodology are:

- development of technique for detection and identification of the impact in advance,
- development of devices for quickly responding, controlled in real time, semi-active detachment of structural members and for introducing initial distortions.

Another challenging field for research is mentioned above development of methodology able to determine optimal control strategies for impact absorption in multi-load cases, when compromise solutions are required.

5 Optimal Adaptation to Impact Load (Large Deformations)

Following the above described procedure for structural remodeling, let us assume that the structural geometry is already determined. Then, the second objective of the smoothest structural adaptation to identified impact can be addressed. Still constraining ourselves to small deformations, the optimal adaptation procedure based on the gradient calculation (7) can be proposed. However, the main contribution to the impact energy dissipation is due to the plastic flow, which makes an enlargement of strokes in controlled dissipaters crucial. In the consequence, large deformations have to be taken into account.

To provide optimal energy absorption it is necessary to perform a process of adaptation, consisting of the following three stages:

- Load identification
- Choosing optimal strategy
- Structural adaptation

The dynamic load level can be evaluated, in advance, before the impact by measuring velocity and estimating the mass of the colliding body. Alternatively , its value might be identified at the beginning of the impact process by sensors embedded into the structure.

In order to dissipate the kinetic energy in an optimal way, one has to apply a correct strategy to the active elements, where the following two strategies are formulated:

- semi-active control,
- active control.

In the first case pre-selected triggering stress levels σ_i in structural elements remain unchanged during impact, in the second one theoretical possibility of real-time changes in control parameters is assumed. Although full

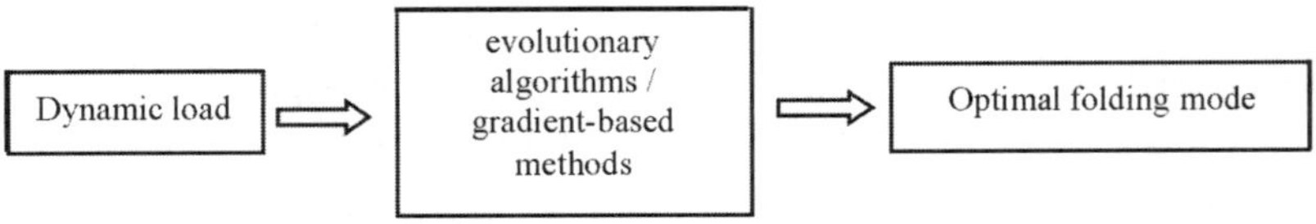

Fig. 8. The algorithm for semi-active control

real-time control does not seem to be feasible in a real design, one can expect that at least a few stress changes might be applied during the impact time.

The problem of optimal adaptation can be formulated as follows:

- Semi-active control – for a given impact load minimize the difference between acceleration values in selected points of the structure and desired response function $\ddot{q}_i^d(t)$:

$$\min f = \sum_t \sum_i \left[\ddot{q}_i(t) - \ddot{q}_i^d(t)\right]^2 \tag{25}$$

- Active control – for a given impact load, for every time step minimize the difference between acceleration values in selected points of the structure and desired response function $\ddot{q}_i^d(t)$:

$$\min f(t) = \sum_i \left[\ddot{q}_i(t) - \ddot{q}_i^d(t)\right]^2 \tag{26}$$

subject to the following constraints

$$|\bar{\sigma}| \in \langle \bar{\sigma}_{\min}, \bar{\sigma}_{\max} \rangle\,, \quad \max\{q\} \leq \bar{q} \tag{27}$$

where $\bar{\sigma}$ denotes the plastic-like yield stress level and $\bar{q}$ is the maximum crush distance.

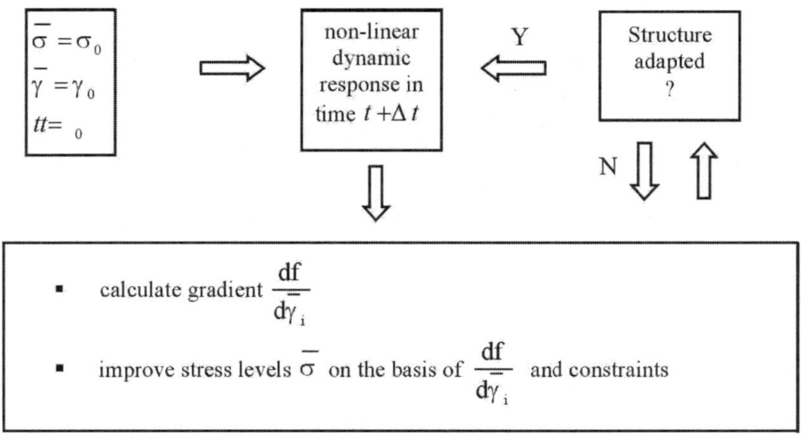

Fig. 9. The algorithm for active control

Satisfactory solution of the above problem exists if the external load intensity is not higher than the maximal safe load level. Beyond this limit a control strategy with the highest possible capability of energy dissipation must be applied.

The Multifolding Microstructure MFM

Let us discuss the truss-like microstructure (similar to honeycomb layout) shown on Fig. 10. Elements arranged in a periodic pattern are equipped with specially designed devices called micro-fuses. Micro-fuses provide control over yield stress in an element. After reaching selected threshold elements exhibit plastic-like behaviour.

Fig. 10. Multifolding microstructure MFM

In order to get the additional value of energy dissipation (due to the synergy of repetitive use of dissipaters) the crucial point is to pre-design the optimal distribution of yield stress levels in all stickers, triggering desired sequence of local collapses.

Two single-column MFM models with a different number of control parameters are presented on Fig. 11. The most basic structure has only two control parameters (Fig. 11(b)): yield thresholds marked 1 and 2. Therefore, only two presented folding sequences are possible.

More complex structure "5" with five control thresholds is depicted in Fig. 11(a). The number of possible folding modes is in this case much greater that in the latter one.

It is clearly visible that different distribution of yield stresses will result in different behaviour of the MFM and different stiffness characteristics. Therefore, structure can adapt itself to the level of dynamic loading.

Because of complex phenomena like plasticity and contact, modeling of the MFM microstructure involves highly nonlinear and numerically expensive dynamic FEM analysis.

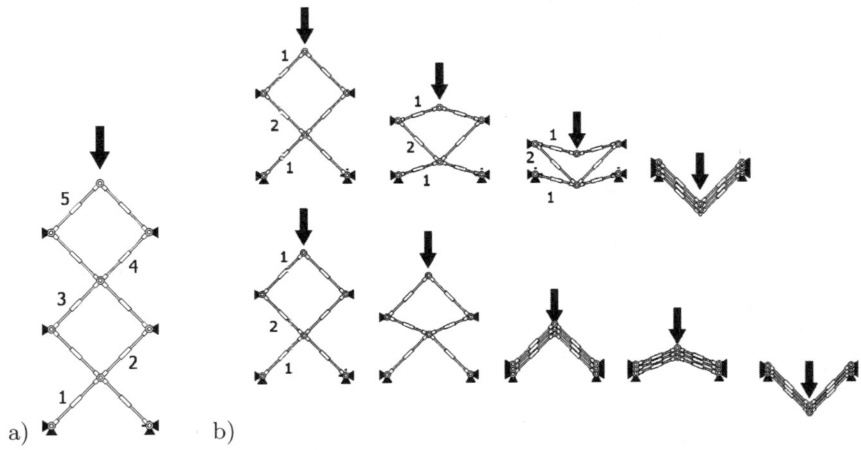

Fig. 11. a) Single column MFM "5" and b) "2" with sample folding sequences

Simplified Model

In order to increase numerical efficiency of the analysis, a simplified analytical model was introduced. The model describes behaviour of a single column of the multifolding microstructure and is based on the following assumptions:

- the material is considered as a rigid-perfectly plastic, therefore the force in a deforming structural member remains constant,
- the mass of the absorber is negligible comparing to the impacting mass,
- the process of deformation is divided into sequences: in each sequence only one level of the microstructure is folding while the other elements remain rigid.
- the dissipated energy related to the final deformation is equal to the initial value of the kinetic energy

Such assumptions allow us to formulate the equilibrium equation based on the second Newton's law for the loaded node, for each sequence:

$$F\left(q(t)\right) = m\ddot{q}\left(q(t)\right) \tag{28}$$

Because the force in the elements, which are active in the current sequence, is constant, the change in the resultant force is dependent only on the change in the geometry of the elements. For the current sequence, this yields:

$$F\left(q\right) = 2\sigma^{\star}_{i,\mathrm{act}} A \sin\theta\left(q\right) = 2\sigma^{\star}_{i,\mathrm{act}} A \frac{h-q}{\sqrt{h^2-\left(h-q\right)^2}}, \quad q \in \langle 0, 2h \rangle \tag{29}$$

Acceleration of the loaded node can be expressed as:

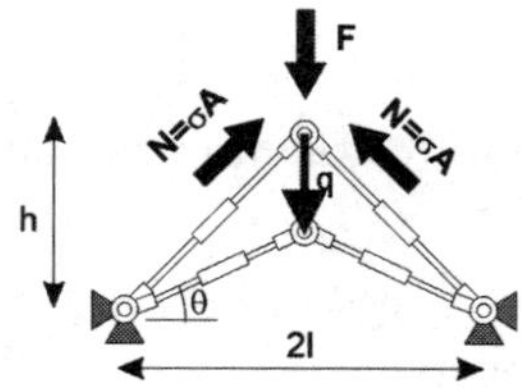

Fig. 12. Kinematics assumed for the simplified model

$$\ddot{q}\,(q) = \frac{F\,(q)}{m}, \quad q \in \langle 0, 2h \rangle \tag{30}$$

Energy of the plastic strain is equal to the work of the resultant force F

$$Dyss^{\mathrm{seq}} = E_{\mathrm{plast}} = \int_{0}^{2h} F\,(q)dq \tag{31}$$

To describe the behaviour of the single-column structure one has to solve the following problem: *for a given set of design parameters* $\boldsymbol{\sigma}$ *and kinetic energy* E_{kin} *find corresponding evolution of acceleration of the loaded node and total energy of the plastic strain* $E_{plast} = \sum_{1}^{Nseq} Dyss^{seq}$.

The initial distribution of the yield stresses $\boldsymbol{\sigma}$ uniquely defines evolution of the deformation. In each sequence elements with the lowest value of yield stress are folding first. The process evolves until the dissipated energy E_{plast} exceeds the initial kinetic energy E_{kin}.

The simplified model provides accuracy in predicting the acceleration and dissipation combined with a very good performance of the algorithm.

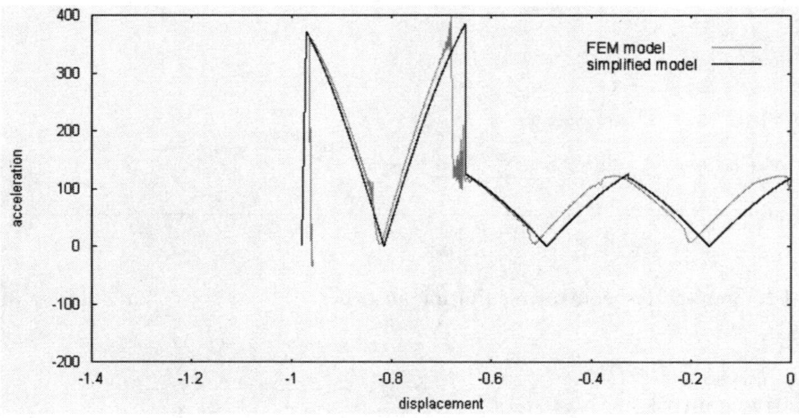

Fig. 13. Results from the simplified and FEM analysis

Control of the MFM

Process of structural adaptation to the impact consists of following stages:

- Load identification
- Selection of optimal strategy of dissipation
- Adaptation of active elements
- Dissipation of the energy

In order to provide optimal results of adaptive impact absorption one has to apply a correct control strategy. Two different approaches, discussed previously in [7], are considered: semi-active and active control. In the first case pre-selected triggering stress levels σ_i in structural elements remain unchanged during an impact, in the second one theoretical possibility of real-time changes in the control parameters is assumed.

- Semi-active control
  The objective function of the semi-active control is to minimize the maximal acceleration of the loaded node: $f = \min(\max_t \{\ddot{q}_{\mathrm{load}}(t)\})$, with constraints imposed on control parameters $\sigma_i \in \langle\sigma_{\min}, \sigma_{\max}\rangle$ and maximal displacement $q_{\mathrm{load}} \leq q_{\max}$.
  Figure 14 presents results obtained for the basic absorber "2" (with and without applied control) compared with results for enhanced absorber "5". It is clearly visible that the multi-layered structure offers much better performance in a very wide range of kinetic energy value (different mass value with constant initial velocity of 15 m/s).

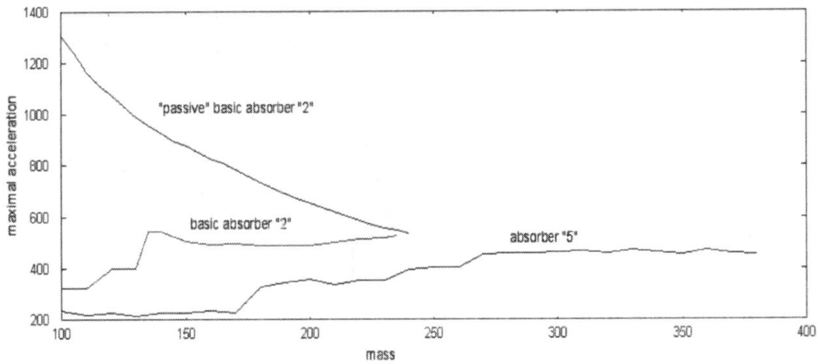

Fig. 14. Semi-active control results for absorbers "2" and "5" for different mass values

- Active control
  Objective function: for every time step, minimize the difference between acceleration in the loaded node and desired response function $\ddot{q}^d$: $\min f(t) = \left[\ddot{q}_{\mathrm{load}}(t) - \ddot{q}^d(t)\right]^2$.

Desired level of acceleration $\ddot{q}^d(t)$ must provide dissipation of kinetic energy of the impact. Therefore, it is necessary to solve an additional optimization task (in this case, based on a simplified approach).
In FEM analysis, at every time step control parameters (hardening coefficient in plasticity model) are updated on the basis of the objective function's gradient, which is calculated by the finite difference approach. A classical algorithm of the steepest gradient descent is applied.
Results, obtained for the basic absorber "2", are presented in Figs. 15 and 16. The strategy of active control is compared to the semi-active and passive one. Active control further improves the performance of the absorber by 20%-30%. Nevertheless, it has to be taken into consideration if such improvement can justify application of a complicated strategy in comparison with the simple and robust semi-active approach.

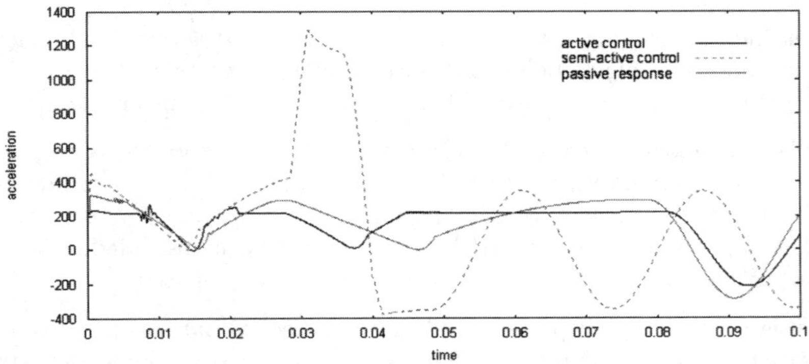

Fig. 15. Results of different control strategies for impacting mass of 100 kg

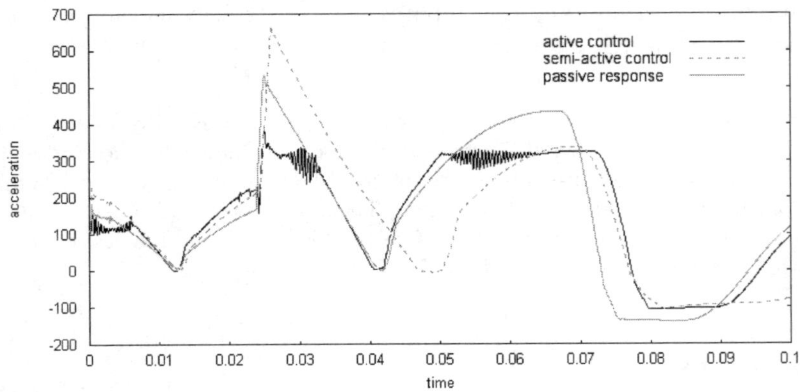

Fig. 16. Results of different control strategies for impacting mass of 200 kg

6 Conclusions

The methodology (based on the so-called Dynamic Virtual Distortion Method) of the design of structures exposed to impact loading is presented in the work. Minimization of material volume and accelerations of structural response are chosen as the objective functions for optimal design of structures adapting to impact loads. The cross-sections of structural members as well as stress levels triggering plastic-like behavior and initial prestressing are the design parameters.

The crucial points in the proposed methodology are:

- development of technique for detection and identification of the impact in advance,
- development of devices for quickly responding, controlled in real time, semi-active detachment of structural members and for introducing initial distortions,
- another challenging field for research is mentioned above development of methodology able to determine optimal control strategies for impact absorption in multi-load cases, when compromise solutions are required.

The paper demonstrates the effectiveness of the proposed concept. The yield stress level adaptation to the applied load has significant influence on the intensity of impact energy dissipation.

The concept of adaptive MFM systems has been discussed, where the following general methodology in the design can be proposed:

- design a topological pattern and material redistribution of the adaptive structure with given initial configuration for variety of all expected extreme loadings,
- particularly, consider the MFM pattern in case of frontal impacts,
- apply in real time the pre-computed control strategy as the response for detected (through a sensor system) impact.

Acknowledgement. This work was supported by grant No. KBN ST07A05222 from the Institute of Fundamental Technological Research funded by the National Research Committee and presents a part of the Ph.D. thesis of the second and the third author supervised by the first author. The authors would also like to gratefully acknowledge the financial support through the 5FP EU project Research Training Networks "SMART SYSTEMS" HPRN-CT-2002-00284.

References

1. Arora JS, Kim CH, Mijar AR (1999) Simplified models for Automotive Crash Simulation and Design Optimization. In: Proceedings of 3rd World Congress of Structural and Multidisciplinary Optimization", Buffalo, New York, USA,. May 17-2

2. Díaz AR, Soto CA (1999) Lattice Models for Crash Resistant Design and Optimization. In: Proceedings of 3rd World Congress of Structural and Multidisciplinary Optimization", Buffalo, New York, USA,. May 17-21
3. Holnicki-Szulc J, Mackiewicz A, Kołakowski P (1998) Design of Adaptive Structures for Improved Load Capacity. AIAA Journal vol.36, No.3
4. Holnicki-Szulc J, Pawlowski P,Wiklo M (2003) High-performance impact absorbing materials - the concept, design tools and applications. Smart Materials and Structures vol.12, number 3
5. Holnicki-Szulc J, Zielinski T, New damage Identification Method Through the Gradient Based Optimisation, Proc. COST International Conference on System Identification & Structural Health Monitoring, Madrid, 6-9 June, 2000.
6. Knap L, Holnicki-Szulc J (1999) Optimal Design of Adaptive Structures for the Best Crash-Worthiness. In: Proc. 3rd World Congress on Structural and Multidisciplinary Optimisation", Buffalo, New York, USA,. May 17-2,
7. Marzec Z Holnicki-Szulc J, Adaptive Barriers with Maximal Impact Energy Absorption, Proc. 3rd World Congress on Structural and Multidisciplinary Optimization, Buffalo, May
8. Maute K, Schwartz S, Ramm E (1998) Adaptive Topology Optimization of Elastoplastic Structures. Structural Optimization 15: 81-89
9. Mayer RR, Kikuchi N, Scott RA (1996) Applications of Topology Optimization Techniques to Structural Crashworthiness. Int. J. Num. Meth. Engrg. 39: 1383-1403
10. Neves MM, Rodrigues H, Guedes JM (1995) Generalized Topology Design of Structures with a Buckling load Criterion. Structural Optimization 10: 71-78
11. Pedersen CBW (2002) Topology Optimization of 2-D Frame Structures with Path Dependent Response. International Journal for Numerical Methods in Engineering
12. Pedersen CBW (2002) Topology Optimization for Crashworthiness of Frame Structures. I: Proceedings of Icrash2002, Society of Automotive Engineering, Melbourne, Australia
13. Yamakawa H, Tsutsui Z, Takemae K, Ujita Y, Suzuki Y (1999) Structural Optimization for Improvement of Train Crashworthiness in Conceptual and Preliminary Designs. In: Proc. 3rd World Congress on Structural and Multidisciplinary Optimisation, Buffalo, New York, USA, May 17-2,
14. Yuge K. Iwai N, Kikuchi N (1998) Topology Optimization Algorithm for Plates and Shells Subjected to Plastic Deformations. In: Proc. 1998 ASME Design Engineering Technical Conference, paper DET98/DAC-5603

Damage Detection of Structures Using Spectral Element Method

Wiesław Ostachowicz[1,2] and Marek Krawczuk[1,3]

[1] Institute of Fluid Flow Machinery, Fiszera 14, 80–952 Gdańsk Poland
[2] Gdynia Maritime University, Faculty of Navigation,
Al. Zjednoczenia 3, 81–345 Gdynia Poland
wieslaw@imp.gda.pl
[3] Technical University of Gdańsk, Faculty of Electrical and Control Engineering,
Narutowicza 11/12 80–952 Gdańsk Poland
mk@imp.gda.pl

1 Introduction

In the paper wave propagation in engineering structures is studied on the basis of detection and estimation of the location, as well as the extent, of structural damage.

Damage detection in systems is based on the known fact that material discontinuities affect the propagation of elastic waves in solids. Elastic waves can be generated and also sensed by an array of transducers either embedded within, or bonded to, the surface of the structure. Piezoceramic material has been selected as the primary candidate for transducer construction. Piezomaterials convert electrical energy to mechanical strain and thus are capable of producing an elastic wave propagating through the structure. These sensors can therefore provide information regarding the condition and damage state of the structures as they age.

Wave frequencies, specifically those that are most sensitive to damage, depend on the type of structure, the type of material, and the type of damage. The main problem in analysing the propagation of high velocity waves in distributed structures is that the spatial discretization must be accurate to capture the amplified effect of wave scattering at structural discontinuities. The conventional modal method, when extended to the high frequency regime, becomes computationally inefficient since many higher modes that participate in the motion will not be represented, whilst adopting the finite element (FE) discretization technique means that computational efficiency decreases dramatically as the system size increases. Among many frequency domain methods, the spectral method is found to be suitable due to its adaptation of the displacement based FE technique and therefore it enables one to handle arbitrary skeletal structures.

SHM (Structural Health Monitoring) technology is the technological platform for a new maintenance philosophy. SHM technology works with a built-in sensor network on a structure. A new philosophy based on SHM technology creates feedback loops within the design, manufacturing, and maintenance procedures by providing additional knowledge about a specific design performance, material quality, and structure condition respectively.

Structural health monitoring, specifically with regard to monitoring damage in highly stressed structures, has become a major issue, particularly in the aerospace environment but also elsewhere. This is mainly driven by the fact that structures are designed for longer life cycles (around 20 to 50 years in the aerospace environment) compared to the different actuation, control and electronic components being replaced in much shorter life cycles and thus making prediction of the operational as well as damaging conditions much more difficult. This has led to the development of structural health monitoring systems that are able to sense and locate damages in structures.

Interest in various non-destructive damage detection methods has considerably increased over the past twenty years. During this time many methods founded on modal analysis techniques have been developed (Adams and Cawley [1], Cawley *et al.* [2], Messina *et al.*[3], Krawczuk and Ostachowicz [4], Lim and Kashangaki [5], Farrar and Jauregui [6]). These techniques are successfully used when monitoring structures where the presence of damage leads to changes in the some of the lower natural frequencies in modal parameters. These changes are often at the same level as the measurement errors. This fact has stimulated recent new methods based on wave propagation models (Lakshmanan and Pines [7], and [8], Chang [9-11, Krawczuk and Ostachowicz [12, 13]). Such models are well suited for the detection of even very small defects given that they are very sensitive to changes in local dynamic impedance.

This paper presents a method of wave propagation which can be further used to detect small delaminations in beam-like and plate structures. The structures are modelled by spectral finite elements.

2 Spectral Finite Elements

The approach is similar in style to that of the finite element method but with the very significant difference that the element stiffness matrix is established in the frequency domain. As a consequence, these spectrally formulated elements describe exactly the wave propagation dynamics, and in contrast to the conventional element this means that elements can span all the way from one point to another without losing fidelity.

Among many frequency domain methods the spectral method is found to be suitable due to its adaptation of displacement based FE technique and therefore it enables one to handle arbitrary skeletal structures. Spectral Element (SE) analysis has been established as a powerful method used for wave propagation. The Spectral Element Method (SEM), a high-accuracy numerical method, combines the accuracy of conventional spectral methods and the geometric flexibility of finite element methods. It is the method of Fourier synthesis (or spectral analysis), where the behaviour of the signal is viewed as a superposition of many infinitely long wave trains of different periods (or frequencies). The actual response is synthesised by a judicious combination of these wave trains. Thus the problem of characterising a signal is transformed into one of determining the set of combination coefficients. These coefficients are called the Fourier transform of the signal. The problem being tackled invariably simplifies when it is expressed in terms of the Fourier transform. The last step in the analysis involves performing an inverse transform (reconstructing the signal).

The SEM approach is based on global approximations of functions, i.e. the analysed functions are at first approximated using basis functions and are then exactly differentiated. As a consequence, relatively small number of elements can be used for modelling without losing accuracy. This is particularly useful for wave propagation modelling.

The basic steps in the development of SEM can be described as follows. The governing wave equations are first transformed from the time domain to the frequency domain using a discrete Fourier transform (DFT). For computational implementation we use the FFT algorithm built into the FE code. The spatial variation is semi-explicitly obtained by solving the characteristic equation in k-space. This results in a complex shape function matrix representing the linear superposition of all the wave modes. Also, the complex dynamic strain–displacement matrix and the dynamically consistent force vector can both be derived. Following the conventional FE method, the complex dynamic stiffness matrix is then formed, and this is exact. This makes the proposed SEM an efficient model suitable for use within the framework of an automated FE method rather than a trade-off for broadband wave propagation analysis. But the fundamental difference from the conventional FE method is that all the spectral amplitudes that correspond to elemental nodal variables are evaluated at each frequency step (FFT sampling points) instead of pseudo-static variables evaluated at each time step or at each eigenfrequency. The global system is solved for unit spectral amplitude of the applied load history at each FFT sampling frequency. Therefore, computation in this way of the frequency response function (FRF) of the global system is straightforward. The time domain response is obtained using the inverse FFT.

In a simple form, the solution to a wave propagation problem can be represented as by Ostachowicz *et al.* [14]:

$$
\begin{aligned}
w = \sum_n \hat{P}_n \left\{ \hat{G}_1 \left(k_{1n} x\right) + \hat{G}_2 \left(k_{2n} x\right) + \ldots \right\} \exp\left(i\omega_n t\right) \\
= \sum_n \hat{P}_n \hat{G}_1 \left(k_{mn} x\right) \exp\left(i\omega_n t\right) \quad (1)
\end{aligned}
$$

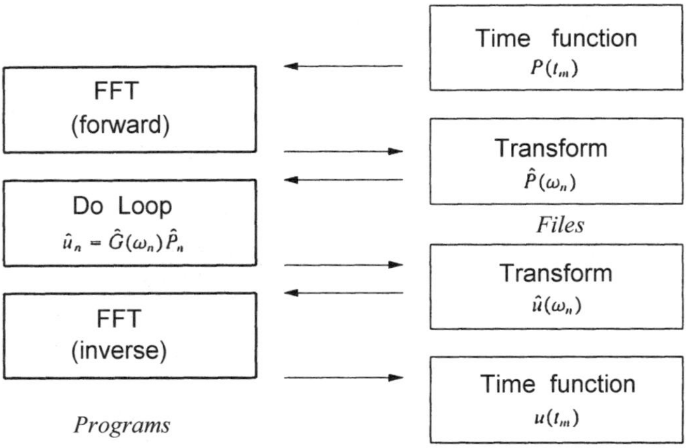

Fig. 1. Flow diagram for the wave reconstruction program

where: $\hat{G}$ is the analytically derived transfer function of position x. $\hat{G}$ has different numerical values at each frequency. $\hat{P}_n$ is the amplitude spectrum and is known from input conditions or from measurements. $\hat{P}_n\hat{G}$ is recognised as the Fourier transform of the solution.

Certainly the solution is different at each position, but once it is evaluated at a particular position then its inverse immediately gives the time history of the solution at that point. Figure 1 shows a flow diagram for the basic algorithm to propagate a wave.

Briefly, the time input $P(t)$ is converted into its spectrum $\hat{P}_n$ through use of the forward FFT. The transformed solution is then obtained by evaluating the product:

$$\hat{u}_n = \hat{P}_n \hat{G}_1 \left(k_{mn} x\right) \tag{2}$$

at each frequency and some position.

This is finally reconstructed in the time domain by use of the inverse FFT. It is necessary to realise (when using the FFT to perform the inversion) that in this process $\hat{P}_n\hat{G}$ is evaluated only up to the Nyquist frequency and the remainder is obtained by imposing the condition that it must be the complex conjugate of the initial part. This ensures that the reconstructed time history is only real.

3 Cracked Rod Spectral Element

A spectral rod finite element with a transverse open and non-propagating crack is presented in Figure 2. The length of the element is L, and its area of cross-section is A. The crack is substituted by a dimensionless spring, which flexibility θ is calculated by using Castigliano's theorem and laws of the fracture mechanics, Krawczuk [15].

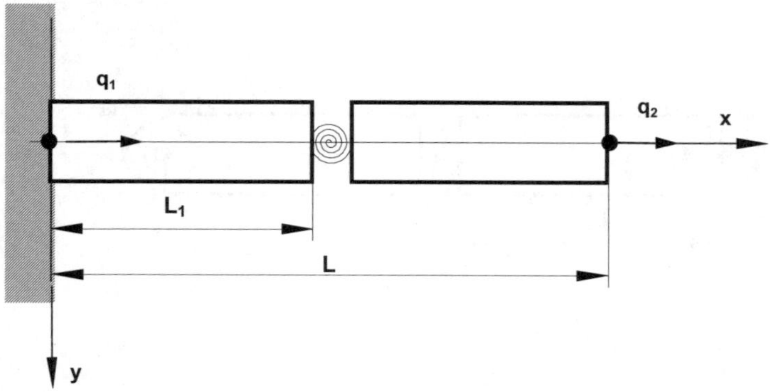

Fig. 2. The model of the rod with a transverse open and not propagating crack simulated by elastic hinge

Nodal spectral displacements are assumed in the following form for the left and right part of the rod:

$$\begin{aligned} \hat{u}_1(x) &= A_1 e^{-ik_n x} + B_1 e^{-ik_n(L_1-x)} && \text{for } x \in (0, L_1) \\ \hat{u}_2(x) &= A_2 e^{-ik_n(x+L_1)} + B_2 e^{-ik_n[L-(L_1+x)]} && \text{for } x \in (0, L-L_1) \end{aligned} \tag{3}$$

where: L_1 denotes the location of the crack, L is the total length of the rod and k_n is the wave number calculated as follows:

$$k_n = \omega_n \sqrt{\frac{\rho}{E}} \tag{4}$$

where: ρ is the density of the rod material, E denotes Young's modulus and ω_n is a natural frequency.

The coefficients A_1, A_2, B_1 and B_2 can be calculated as a function of the nodal spectral displacements using the element boundary conditions (see Krawczuk and Ostachowicz [16]). Using boundary conditions can relate the coefficients A_1, A_2, B_1 and B_2 to the nodal spectral displacements. The nodal spectral forces can be determined by differentiating the spectral displacements with respect to x. Finally, using the relation between nodal forces and nodal displacements the dynamic stiffness matrix of a cracked rod spectral finite element can be evaluated.

4 Cracked Timoshenko Beam Spectral Element

A spectral Timoshenko beam finite element with a transverse open and non-propagating crack is presented in Figure 3. The length of the element is L, and its area of cross-section is A. The crack is substituted by a dimensionless and massless spring, whose bending θ_b and shear θ_s flexibilities are calculated using Castigliano's theorem and the laws of fracture mechanics, Krawczuk [15].

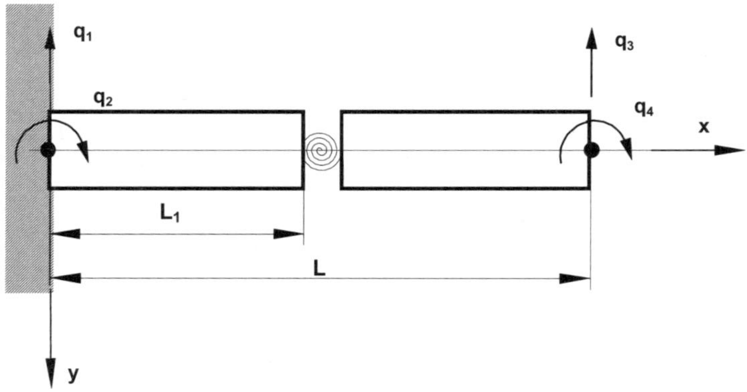

Fig. 3. The model of the Timoshenko beam with a transverse open and not propagating crack simulated by elastic hinge

Nodal spectral displacements $\hat{w}$ and rotations $\hat{\phi}$ are assumed in the following forms, for the left and right part of the Timoshenko beam:

$$\begin{aligned}
\hat{w}_1 =& R_1 A_1 \mathrm{e}^{-ik_1 x} + R_2 B_1 \mathrm{e}^{-ik_2 x} - R_1 C_1 \mathrm{e}^{-ik_1 (L_1 - x)} - R_2 D_1 \mathrm{e}^{-ik_2 (L_1 - x)} \\
& \text{for } x \in (0, L_1) \\
\hat{\phi}_1 =& A_1 \mathrm{e}^{-ik_1 x} + B_1 \mathrm{e}^{-ik_2 x} + C_1 \mathrm{e}^{-ik_1 (L_1 - x)} + D_1 \mathrm{e}^{-ik_2 (L_1 - x)} \\
& \text{for } x \in (0, L_1) \\
\hat{w}_2 =& R_1 A_2 \mathrm{e}^{-ik_1 (x + L_1)} + R_2 B_2 \mathrm{e}^{-ik_2 (x + L_1)} - R_1 C_2 \mathrm{e}^{-ik_1 (L - (L_1 + x))} \\
& - R_2 D_2 \mathrm{e}^{-ik_1 (L - (L_1 + x))} \quad \text{for } x \in (0, L - L_1) \\
\hat{\phi}_2 =& A_2 \mathrm{e}^{-ik_1 (x + L_1)} + B_2 \mathrm{e}^{-ik_2 (x + L_1)} + C_2 \mathrm{e}^{-ik_1 (L - (L_1 + x))} \\
& + D_2 \mathrm{e}^{-ik_1 (L - (L_1 + x))} \quad \text{for } x \in (0, L - L_1)
\end{aligned} \tag{5}$$

where: L_1 denotes the location of the crack, L is the total length of the beam, R_n is the amplitude ratios given by:

$$R_n = \frac{ik_n GAK_1}{GAK_1 k_n^2 - \rho A \omega^2} \text{ for } (n = 1, 2) \tag{6}$$

where: $K_1 = \left(\frac{0.87+1.12\nu}{1+\nu}\right)^2$, ν is Poisson ratio, G is shear modulus, ρ denotes density of the material and ω is a natural frequency.

The wave numbers k_1 and k_2 are roots of characteristic equation in the general form:

$$(GAK_1 EJ)k^4 - (GAK_1 \rho J K_2 \omega^2 + EJ\rho A\omega^2)k^2 + \left(\rho J K_2 \omega^2 - GAK_1\right) \rho A \omega^2 = 0 \tag{7}$$

where: $K_2 = 12K_1/\pi^2$, E denotes Young's modulus and J is the geometrical moment of inertia of the beam cross-section.

The coefficients A_1, B_1, C_1, D_1, A_2, B_2, C_2 and D_2 can be calculated as a function of the nodal spectral displacements using the boundary conditions, as in Krawczuk and Ostachowicz [17]. The nodal spectral forces ($\hat{T}$ – shear force, $\hat{M}$ – bending moment) can be determined by differentiating the spectral displacements with respect to x. Finally, using relation between the nodal forces and nodal displacements the dynamic stiffness matrix of a cracked rod spectral finite element can be evaluated.

5 Delaminated Multilayer Composite Beam Spectral Element

A schematic of a conceptual damage model of a composite beam is presented in Figure 4.

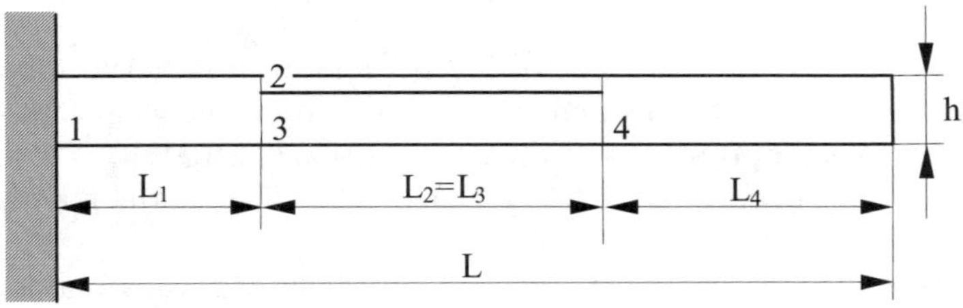

Fig. 4. Model of a composite beam with delamination

The beam under consideration is divided into four parts with lengths $L_1, L_2 = L_3$ and L_4. The length of the beam is L, width b, and height h. The delamination is situated between parts 2 and 3. Its length is $L_2 = L_3$. Displacements and rotations are assumed with Timoshenko beam theory and have the following form:

$$\begin{aligned}
\hat{w}_1(x) =& R_{1,1}A_1\mathrm{e}^{-\mathrm{i}k_1x} + R_{2,1}A_2\mathrm{e}^{-\mathrm{i}k_2x} - R_{1,1}A_3\mathrm{e}^{-\mathrm{i}k_1(L_1-x)} \\
&- R_{2,1}A_4\mathrm{e}^{-\mathrm{i}k_2(L_1-x)} \quad \text{for } x \in (0, L_1) \\
\hat{\phi}_1(x) =& A_1\mathrm{e}^{-\mathrm{i}k_1x} + A_2\mathrm{e}^{-\mathrm{i}k_2x} + A_3\mathrm{e}^{-\mathrm{i}k_1(L_1-x)} \\
&+ A_4\mathrm{e}^{-\mathrm{i}k_2(L_1-x)} \quad \text{for } \in (0, L_1) \\
\hat{w}_2(x) =& R_{1,2}A_5\mathrm{e}^{-\mathrm{i}k_1x} + R_{2,2}A_6\mathrm{e}^{-\mathrm{i}k_2x} - R_{1,2}A_7\mathrm{e}^{-\mathrm{i}k_1(L_2-x)} \\
&- R_{2,2}A_8\mathrm{e}^{-\mathrm{i}k_2(L_2-x)} \quad \text{for } x \in (L_1, L_1 + L_2) \\
\hat{\phi}_2(x) =& A_5\mathrm{e}^{-\mathrm{i}k_1x} + A_6\mathrm{e}^{-\mathrm{i}k_2x} + A_7\mathrm{e}^{-\mathrm{i}k_1(L_2-x)} \\
&+ A_8\mathrm{e}^{-\mathrm{i}k_2(L_2-x)} \quad \text{for } x \in (L_1, L_1 + L_2)
\end{aligned} \tag{8}$$

$$
\begin{aligned}
\hat{w}_3(x) =& R_{1,3}A_9\mathrm{e}^{-\mathrm{i}k_1x} + R_{2,3}A_{10}\mathrm{e}^{-\mathrm{i}k_2x} - R_{1,3}A_{11}\mathrm{e}^{-\mathrm{i}k_1(L_2-x)} \\
& - R_{2,3}A_{12}\mathrm{e}^{-\mathrm{i}k_2(L_2-x)} \quad \text{for } x \in (L_1, L_1+L_3) \\
\hat{\phi}_3(x) =& A_9\mathrm{e}^{-\mathrm{i}k_1x} + A_{10}\mathrm{e}^{-\mathrm{i}k_2x} + A_{11}\mathrm{e}^{-\mathrm{i}k_1(L_2-x)} \\
& + A_{12}\mathrm{e}^{-\mathrm{i}k_2(L_2-x)} \quad \text{for } x \in (L_1, L_1+L_3) \\
\hat{w}_4(x) =& R_{1,4}A_{13}\mathrm{e}^{-\mathrm{i}k_1x} + R_{2,4}A_{14}\mathrm{e}^{-\mathrm{i}k_2x} - R_{1,4}A_{15}\mathrm{e}^{-\mathrm{i}k_1(L-x)} \\
& - R_{2,4}A_{16}\mathrm{e}^{-\mathrm{i}k_2(L-x)} \quad \text{for } x \in (L_1+L_3, L) \\
\hat{\phi}_4(x) =& A_{13}\mathrm{e}^{-\mathrm{i}k_1x} + A_{14}\mathrm{e}^{-\mathrm{i}k_2x} + A_{15}\mathrm{e}^{-\mathrm{i}k_1(L-x)} \\
& + A_{16}\mathrm{e}^{-\mathrm{i}k_2(L-x)} \quad \text{for } x \in (L_1+L_3, L)
\end{aligned}
\tag{8}
$$

where: $L_{1,2}$ denote the beginning and the end of the delamination area, R_n are the amplitude ratios given by Doyle [18]:

$$
R_{n,j} = \frac{\mathrm{i}k_n bA_{66,j}I_{o,j}}{bA_{66,j}I_{o,j}k_n^2 - I_{2,j}\omega^2} \quad \text{for } n = 1, 2; \;\; j = 1 \text{ to } 4 \tag{9}
$$

where: A_{66} and D_{11} are stiffness coefficients and are functions of individual ply properties and orientation, and are integrated over the beam cross-section, I_o and I_2 are inertia properties of the cross-section – note that ω is a natural frequency and i is defined as $\mathrm{i} = \sqrt{-1}$. Subscript j denotes the number of the part of the beam element.

The coefficients A_j ($j = 1$ to 16) can be calculated as a function of the nodal spectral displacements, taking into account the boundary conditions at the tips of the delamination (see Krawczuk and Ostachowicz [17]).

The nodal spectral forces can be determined by differentiating the spectral displacements with respect to x. The nodal spectral forces for the left hand side end of the beam ($x_1 = 0$) are formulated as follows:

$$
\hat{M}_1(x_1 = 0) = D_{11,1}\frac{\partial^2 \hat{w}_1(x_1)}{\partial x_1^2} \tag{10}
$$

$$
\hat{T}_1(x_1 = 0) = D_{11,1}\frac{\partial^3 \hat{w}_1(x_1)}{\partial x_1^3} - I_{2,1}\omega\frac{\partial \hat{w}_1(x_1)}{\partial x_1} \tag{11}
$$

and the nodal spectral forces for the right hand side end of the beam ($x_4 = L_4$) are formulated as follows:

$$
\hat{M}_2(x_4 = L) = D_{11,4}\frac{\partial^2 \hat{w}_4(x_4)}{\partial x_4^2} \tag{12}
$$

$$
\hat{T}_2(x_4 = L) = D_{11,4}\frac{\partial^3 \hat{w}_4(x_4)}{\partial x_4^3} - I_{2,4}\omega\frac{\partial \hat{w}_4(x_4)}{\partial x_4} \tag{13}
$$

Then the formulas (10–13) can be expressed in matrix form:

$$\left\{\hat{P}\right\} = \left[B_{(4\times16)}\right] \cdot \{A\} \qquad (14)$$

The coefficients A_j ($j = 1$ to 16) in Equation (14) are a function of the nodal spectral displacements, and the frequency dependent dynamic stiffness matrix, which correlate the nodal spectral forces with the nodal spectral displacements to be formulated as follows:

$$\left\{\hat{P}\right\} = \left[\hat{K}\left(\omega_n\right)\right]\{\hat{q}\} \qquad (15)$$

where, $\left\{\hat{P}\right\} = \mathrm{col}\left[\hat{M}_1, \hat{T}_1, \hat{M}_2, \hat{T}_2\right]$ and $\{\hat{q}\} = \mathrm{col}\left[\hat{q}_1,\ \hat{q}_2,\ \hat{q}_3,\ \hat{q}_4\right]$.

6 Spectral Plate Element with a Crack

A model of a spectral plate finite element with a transverse open and non-propagating crack is presented in Figure 5. The length of the element in the x direction is L, in the y direction is W, and the thickness of the plate is H. The crack is located at a distance of L_1 from the left edge of the plate and has length $2c$.

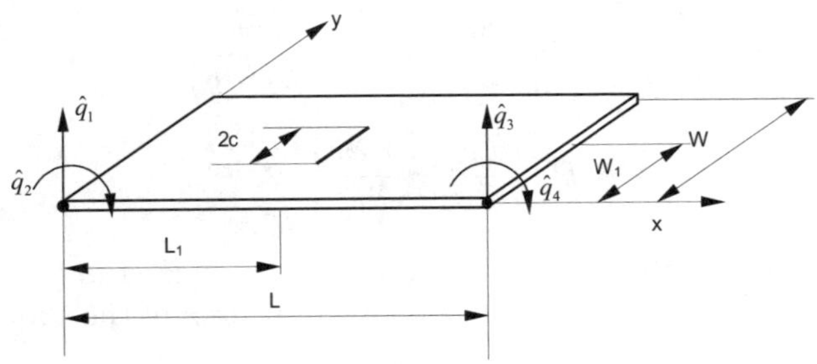

Fig. 5. The model of the plate with a transverse open and non-propagating crack

Nodal spectral displacements (dependent on the wave numbers, consequently on frequencies) are assumed in the following form, for the left and right part of the plate:

$$\hat{w}_1(x,y) = \sum_{n=1}^{N}\sum_{m=1}^{M}\Big(A_{mn}e^{-ik_1x} + B_{mn}e^{-ik_2x} + C_{mn}e^{-ik_1(L_1-x)}$$

$$+D_{mn}e^{-ik_2(L_1-x)}\Big)\,e^{-i\xi_m y} \quad \text{for } x \in (0, L_1),\ y \in (0, W) \qquad (16)$$

$$w_2(x,y) = \sum_{n=1}^{N}\sum_{m=1}^{M}\Big(E_{mn}e^{-ik_1(L_1-x)} + F_{mn}e^{-ik_2(L_1-x)}$$

$$+G_{mn}e^{-ik_1(L-(L_1+x))} + H_{mn}e^{-ik_2(L-(L_1+x))}\Big)\,e^{-i\xi_m y}$$

$$\text{for } x \in (0, L-L_1),\ y \in (0, W) \quad (17)$$

where: L_1 denotes the location of the crack, L is the total length of the plate in the x direction and $k_{1,2}$ are the waves numbers calculated as follows:

$$k_1 = \sqrt{\beta^2 - \xi^2}, \quad k_2 = -\mathrm{i}\sqrt{\beta^2 + \xi^2} \quad (18)$$

with:

$$\xi_m = \frac{2\pi m}{W}, \quad \beta^4 = \frac{\rho h\omega_n^2 - \mathrm{i}\eta H\omega_n}{D}, \quad D = \frac{EH^3}{12\,(1-\nu^2)}, \quad (19)$$

where: ρ is the density of the plate material, η is the damping factor, D is the plate stiffness, ν is the Poisson ratio, E denotes Young's modulus and ω_n is a natural frequency. The coefficients A_{mn}, B_{mn}, C_{mn}, D_{mn}, E_{mn}, F_{mn}, G_{mn} and H_{mn} can be calculated as a function of the nodal spectral displacements using the element boundary conditions (Krawczuk *et al.*, [19]).

The nodal spectral forces can be determined by differentiating the spectral displacements with respect to x and y in the form:

$$\hat{T} = D\left[\frac{\partial^3\hat{w}}{\partial x^3} + (2-\nu)\,\frac{\partial^3\hat{w}}{\partial x^2\partial y}\right] \quad (20)$$

$$\hat{M} = D\left[\frac{\partial^2\hat{w}}{\partial x^2} + \nu\frac{\partial 2\hat{w}}{\partial y^2}\right] \quad (21)$$

From the commonly known fact that the dynamic stiffness matrix can relate forces with displacements, after taking into account the above relations, forces (20) and (21) can be rewritten as:

$$\begin{bmatrix}\hat{T}(0,y)\\ \hat{M}(0,y)\\ \hat{T}(L,y)\\ \hat{M}(L,y)\end{bmatrix} = [K]\cdot\begin{bmatrix}\hat{q}_1\\ \hat{q}_2\\ 0\\ \vdots\\ 0\\ \hat{q}_3\\ \hat{q}_4\end{bmatrix} \quad (22)$$

where:

$$[K] = \begin{bmatrix}P_1 & 0\\ 0 & P_2\end{bmatrix}\begin{bmatrix}M_1 & 0\\ M_2 & M_3\\ 0 & M_4\end{bmatrix}^{-1} \quad (23)$$

where: the square matrix $[K]$ denotes the frequency dependent dynamic stiffness matrix for the spectral plate element with a transverse open and non-propagating crack.

The dimensionless form of the bending flexibility at both sides of the crack can be expressed using formulas elaborated by Khadem and Rezee [20], as follows:

$$\theta\left(\frac{\bar{y}}{W}\right) = \frac{6H}{W}\alpha_{bb}\left(\frac{\bar{y}}{W}\right) \cdot F\left(\frac{\bar{y}}{W}\right) \tag{24}$$

where: h is the thickness of the plate, W is the width of the plate, $\alpha_{bb}\left(\frac{\bar{y}}{W}\right)$ is a function representing the dimensionless bending compliance coefficient and $F\left(\frac{\bar{y}}{W}\right)$ is a correction function. The function $\alpha_{bb}\left(\frac{\bar{y}}{W}\right)$ is given in [19].

7 Numerical Examples

7.1 Longitudinal Wave Propagation in Cracked Rod

In the aim to demonstrate the validity of the proposed model numerical tests have been carried out. All numerical tests were carried out for a cantilever

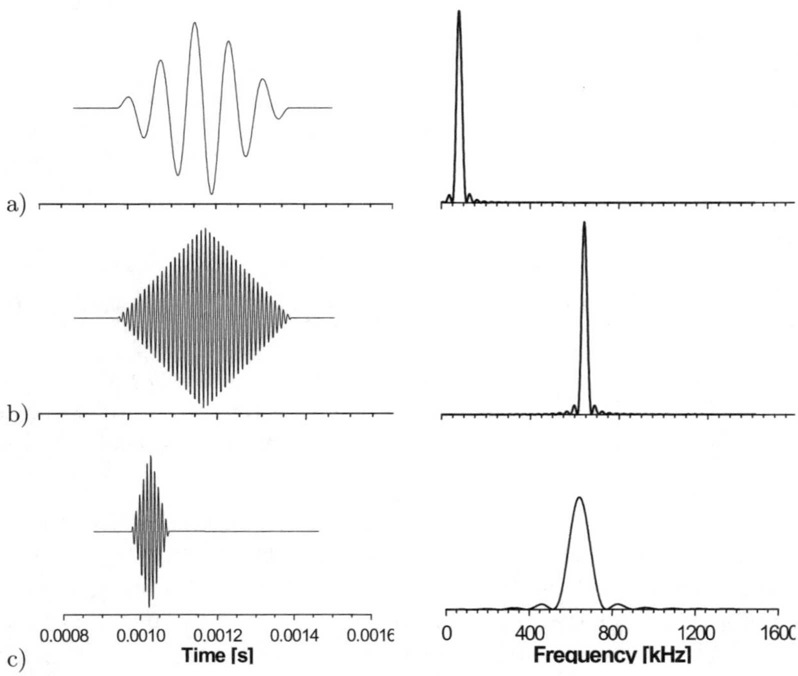

Fig. 6. The tested signals and their FFT, a) signal containing 10 sinusoids per packet lasting for 0,39 μs, b) signal containing 80 sinusoids per packet lasting for 0,39 μs, c) signal containing 20 sinusoids per packet lasting for 0,0925 μs

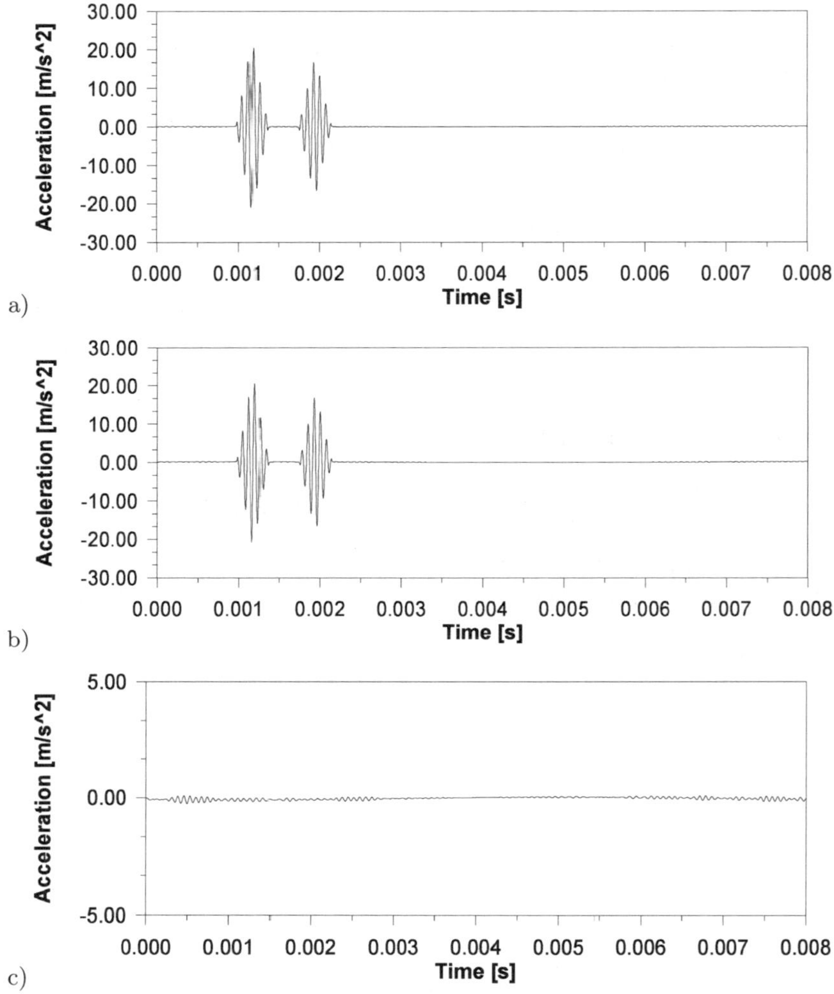

Fig. 7. Acceleration at the second node in a rod, a) rod uncracked, b) rod cracked, the depth of the crack 5%, located in the middle of the rod, c) difference in the signals a and b. Excitation force with 10 sinusoids per packet acting for 0,39 μs

rod with dimensions as follows: the length 2 m, the height 0,02 m, the width 0,02 m, the Young's modulus 210 GPa, and mass density 7860 kg/m^3. The numerical examples show the influence of kind of the excitation signal on the wave propagation. The rod was excited at the free end to vibrations by modulated signals (called packets) – in the examples different sinusoid functions multiplied by a triangle were used. Such a product gives a convolution in the frequency domain. The convolution tends to distribute the effect of a pulse. Figure 6 shows three tested signals and also their FFT to show their run in a frequency domain. As for the signals tested it can be noticed that signals

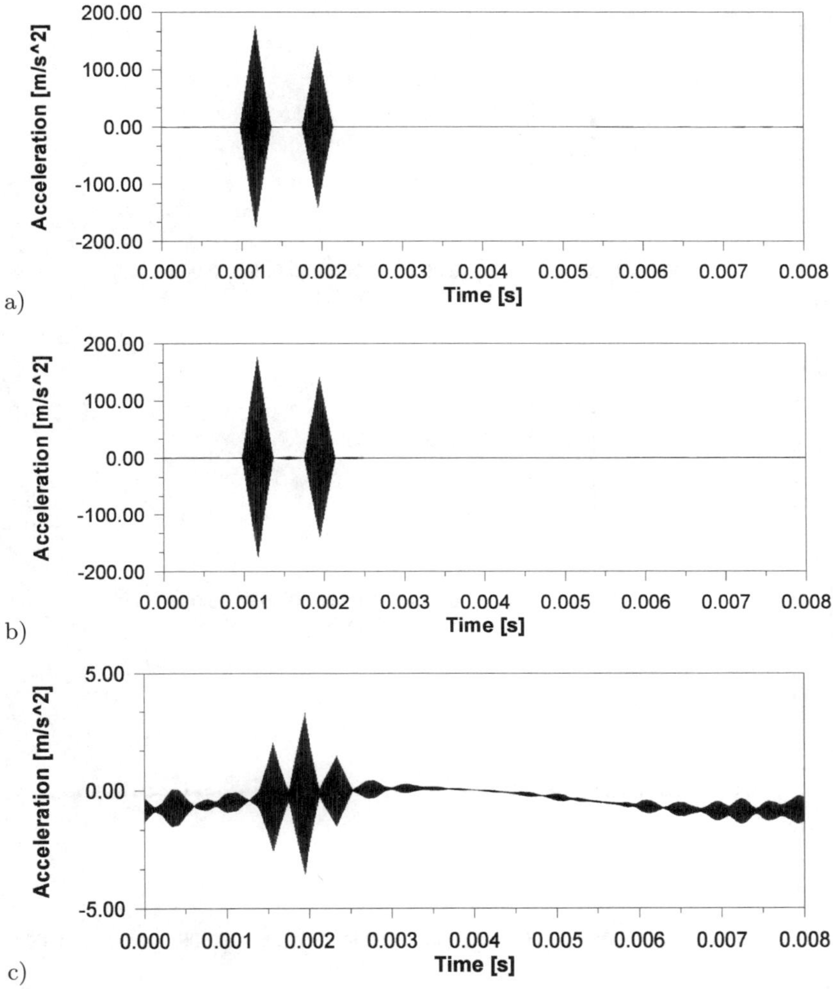

Fig. 8. Acceleration at the second node in a rod, a) rod uncracked, b) rod cracked, the depth of the crack 5%, located in the middle of the rod, c) difference in the signals a and b. Excitation force with 80 sinusoids per packet acting for 0,39 μs

which last for a shorter period of time give better pulse effect in a frequency domain. Such impact forces are practically used as an excitation signals.

For the damage detection process it is extremely important to be able to record the changes between the signals obtained from the damaged and undamaged element. Figures from 7 to 9 show different responses obtained from different excitation forces. The influence of the measurement error was examined in every numerical test done. The answer measured was containing a random noise of a maximum level of 2%.

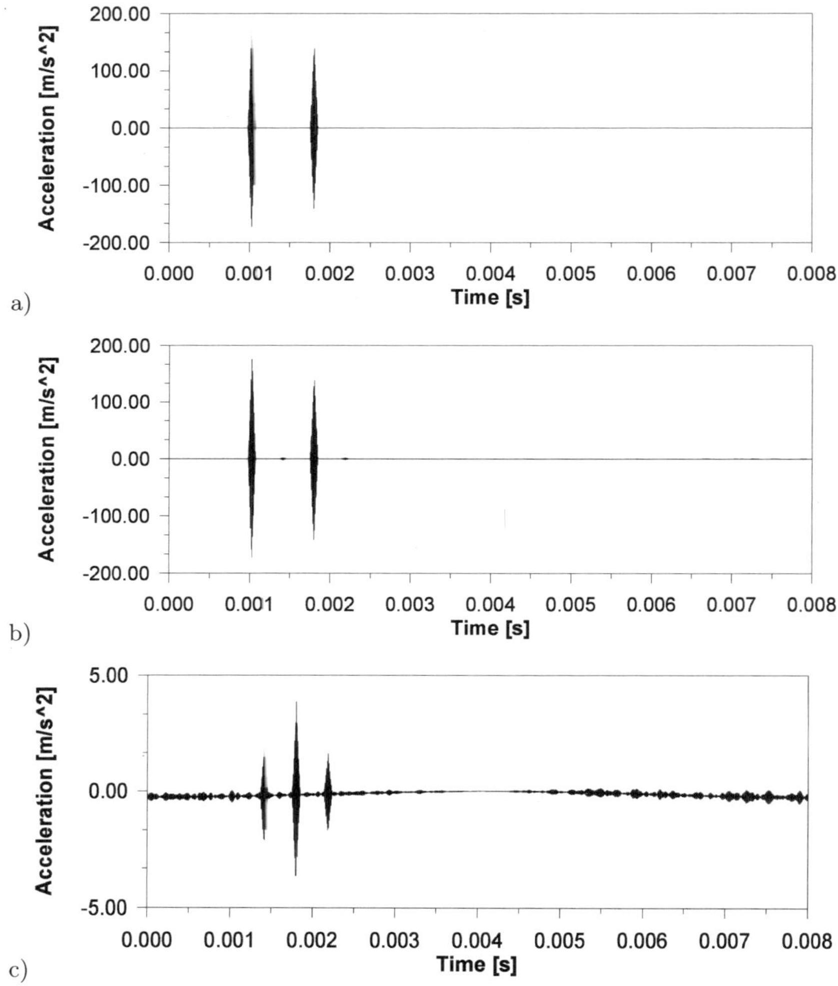

Fig. 9. Acceleration at the second node in a rod, a) rod uncracked, b) rod cracked, the depth of the crack 5%, located in the middle of the rod, c) difference in the signals a and b. Excitation force with 20 sinusoids per packet acting for 0,0925 μs

In the examples maintained there are acceleration responses at the impacted node shown. The meaning of the figures marked as a) is that initial pulse causes a forward – moving tensile one. This latter pulse, on reflection from the fixed end, retains its sign and is picked up later as a compressive pulse (with negative particle acceleration) at the impacted node. Figures marked b) show similar wave propagation in the time domain, but for a rod with a 5% crack. Figures marked with c) show the differences between signals from damaged and undamaged rods. It can be clearly observed that the best results are obtained for the signal, which lasts for a very short period of time (0,0925 μs) and contains 20 sinusoids per packet. With certain and not complicated simple mathematical transform the location of the crack can be obtained from the difference of those signals. It can be also easily noticed that better results are obtained for shorter excitation signals.

7.2 Flexural Wave Propagation in Cracked Beam

In the aim to demonstrate the validity of the proposed model, numerical tests have been carried out for a cantilever Timoshenko beam with dimensions: length 2 m, height 0,02 m, width 0,02 m, Young's modulus 210 GPa, and mass density 7860 kg/m^3. In the numerical tests, the model was fixed at one end and impacted at the second. The Timoshenko beam model consisted of two elements, one with a crack and the second one with a throw – off element. This approach allowed to obtain a moving forward wave and also the backward moving wave.

The numerical example shows the influence of location of the crack on the system response. Figures 10 to 12 show the propagation of a wave in a cracked Timoshenko beam. The excitation force was the packet of 20 sinusoids acting for 0,0925 μs. The location of the crack was changed to find out how the model will behave when the location of the crack will move away from the excitation node. It can be concluded that if the crack is located closer to the

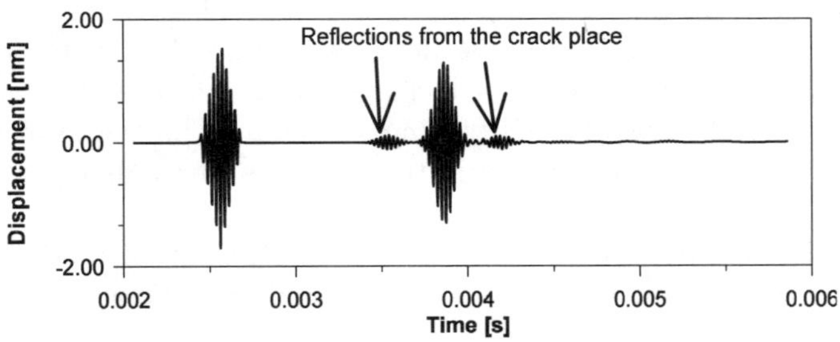

Fig. 10. Wave propagation, for a crack depth of 20% of a beam height and at distance 25% of the length from the fixed end

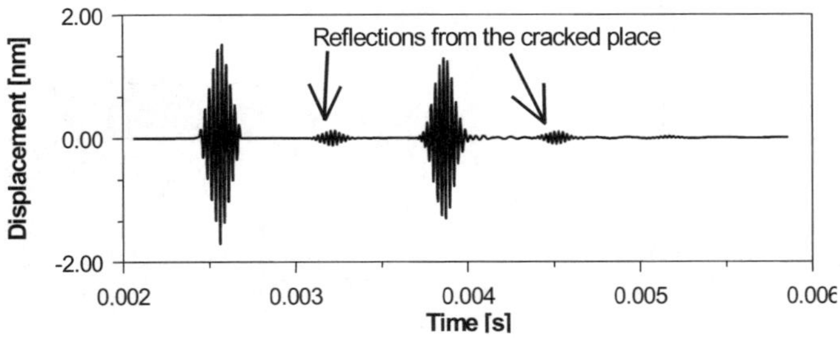

Fig. 11. Wave propagation, for a crack depth of 20% of a beam height and at distance 50% of the length from the fixed end

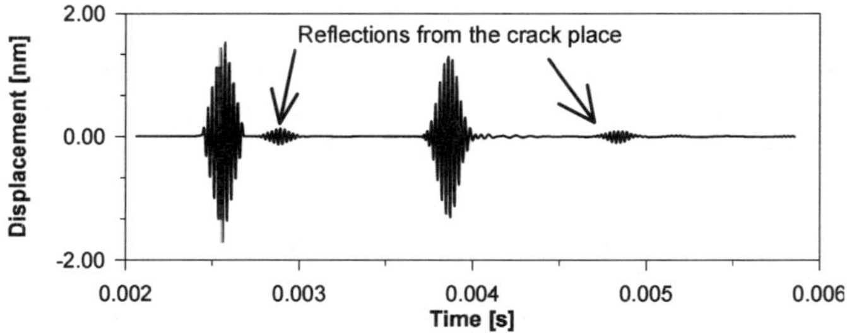

Fig. 12. Wave propagation, for a crack depth of 20% of a beam height and at distance 75% of the length from the fixed end

node the back moving wave is superposing the forward moving wave. If the crack is smaller, returning waves have smaller amplitudes. When the depth of the crack falls down, the number of reflected waves reduces.

7.3 Coupled Flexural-Shear Wave Propagation in Delaminated Composite Beam

All numerical calculations were done for 12 layer cantilever beam made out of glass – epoxy composite. The properties of the composite material used in numerical calculations were as: modules of elasticity: $E_m = 3{,}43\,\mathrm{GPa}$, $E_f = 66{,}5\,\mathrm{GPa}$, Poisson ratios: $\nu_m = 0{,}35$, $\nu_f = 0{,}23$, modules of rigidity: $G_m = 1{,}27\,\mathrm{GPa}$, $G_f = 27\,\mathrm{GPa}$, mass densities: $\rho_m = 1250\,\mathrm{kg/m^3}$, $\rho_f = 2250$ $\mathrm{kg/m^3}$, (lower index m denotes matrix and lower index f denotes fibres of composite material). Each layer's thickness is 1 mm. The length of the beam is 2 m, the width is 0,02 m. The excitation signal is the same as presented in Point 7.2.

Numerical example was done in order to analyse the influence of size of delamination located along neutral axis on wave propagation. Figures 13–15

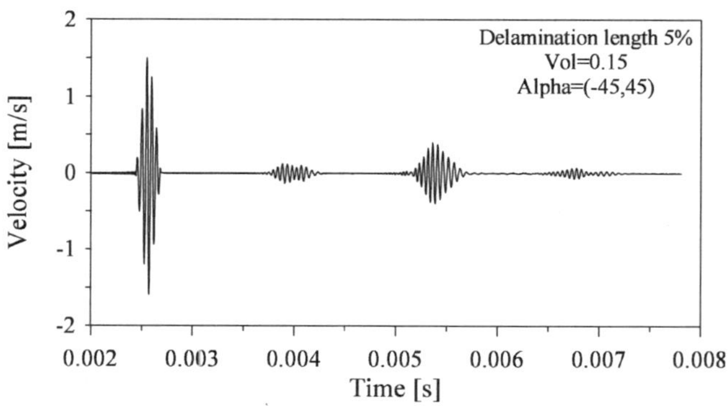

Fig. 13. Delamination length equal 5%, located along neutral axis, centre of delamination located in the 1 m from the fixed end

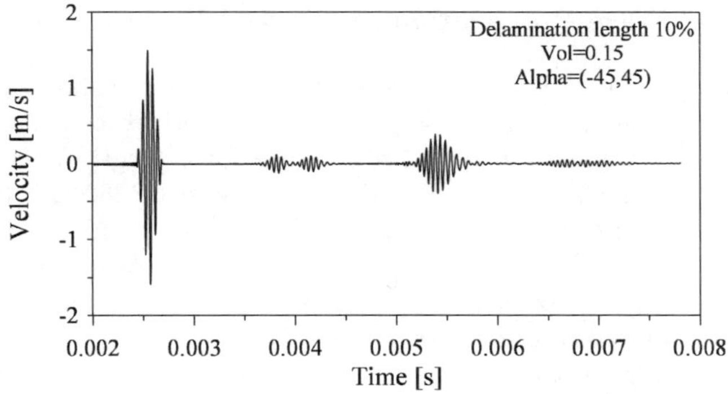

Fig. 14. Delamination length equal 10%, located along neutral axis, centre of delamination located in the 1 m from the fixed end

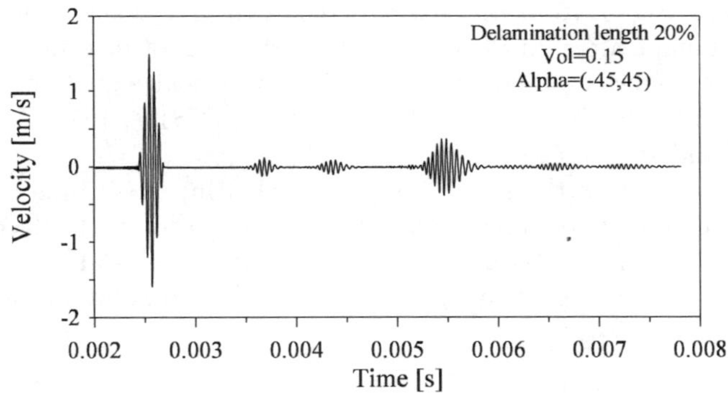

Fig. 15. Delamination length equal 20%, located along neutral axis, centre of delamination located in the 1 m from the fixed end

illustrate responses recorded at the free end of the beam. For this example it was assumed that the ply angles of the composite fibres were (−45/45) degrees and the volume fraction of fibres was equal to 15%. Three different delamination sizes were introduced. Figure 13 illustrates changes for the smallest delamination size, that is 5% of the beam length. Figure 14 is devoted to the delamination which was equal to 10% of the total beam length and Figure 15 presents results for the delamination with length equal to 20% of the total beam length. Additional signal appears between the excitation and the reflection from the fixed end. That is reflection from the delamination edges. It can be observed that when the delamination dimensions grow, reflections from the first and second tip of delamination appear. Significant for this numerical example is fact, that even relatively very small delamination with size equal to 5% of the beam length gives additional reflections. That proves the efficiency of wave propagation based damage detection systems.

7.4 Flexural Wave Propagation in Cracked Plate

To demonstrate the process of wave propagation with elaborated model of a cracked plate exemplary result is presented. Numerical tests have been carried out for a clamped-free plate with dimensions: length 5 m, width 5 m, height 0,05 m, Young's modulus 210 GPa, mass density 7860 kg/m^3, Poisson's ratio 0,33, damping factor 5. The crack was located in the middle of the length of the plate and was parallel to the width edge. The plate was excited with the loads calculated as presented in example 7.2 and the signal was recorded in the middle point of the free end of the plate.

Elaborated numerical example presents changes in propagating wave in a cracked plate. Figures from Fig. 16 present waves caught in time samples starting at 2,44 ms from the beginning of measurement and ending at 3,528 ms, every 24,4 μs. Example presents only several signals measured in subsequent time samples. Signal from Fig. 16a was recorded at 2,44 ms, following signal (16b) was caught after 2,673 ms from the beginning of measurement. Signal from picture 16c was measured at 2,795 ms. And next signals were recorded as follows: 16d – 2,917 ms, 16e – 3,039 ms, 16f – 2,978 ms, 16g – 3,284 ms, 16h – 3,406 ms and 16i – 3,528 ms.

It may be seen on the picture when propagating wave runs to the crack place, divides into two signals. One reflects from the crack place and goes back to the free end, and the second goes to the fixed end. Then the wave reflects from the fixed end propagates to the excitation point, however when it

a) d) g) b) e) h) c) f) i)

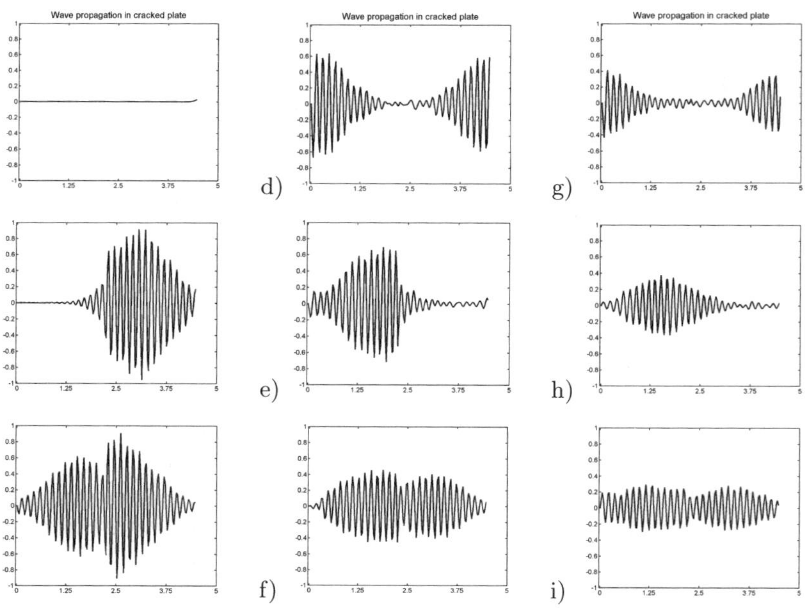

Fig. 16. Wave propagation in a plate with a 5% crack

meets the crack place again divides into two waves. The process continues till wave completely extinguishes. Wave reflections obtained from a crack place are visible as additional peaks in the signal time history. From the information about the exact time of reflection and the velocity of the wave distance, and the same damage localisation may be calculated.

8 Conclusions

This paper has presented and discussed the dynamics of a cracked rod, a cracked Timoshenko beam, a delaminated multilayer composite beam, and cracked plate spectral finite elements. The basic difference between the classical approach and the spectral element method is clearly shown. The way of modelling the stiffness loss due to the crack appearance and the excitation force has also been presented. The investigated models can be used for the analysis of wave propagation. It is easily seen that spectral approach is sufficient for damage detection.

From the numerical examples it arises that the appearance of the crack causes a different time history for the signal compared with the time history of the signal measured from an undamaged structure. The results obtained indicate that the current approach is capable of detecting cracks and delaminations of very small size, even in the presence of considerable measurement errors.

Only from the differences between the signals measured from undamaged and damaged structures can there be information about the location of the crack. The influence of the growth of the damage for wave propagation in the damaged structure has been shown.

The approach presented is very promising as a fatigue damage detection method. As concluded in papers by Krawczuk *et al.* [21, 22] spectral analysis is very sensitive and allows one to detect damage in its very early state of growth. This fact is extremely important from the practical and economical point of view.

References

1. Adams RD, Cawley P (1979) The localisation of defects in structures from measurements of natural frequencies. J. Strain Analysis 14(2): 49-57
2. Cawley P, Adams RD, Pye CJ, Stone BJ (1978) A vibration technique for non-destructively assessing the integrity of structures. J. Mech. Eng. Sci. 20(2): 93-100
3. Messina I, Jones A, Williams EJ (1992) Damage detection and localisation using natural frequency changes. Proc. 1st Conference on Structural Identification, Cambridge, 67-76
4. Krawczuk M, Ostachowicz W (1996) Damage indicators for diagnostic of fatigue cracks in structures by vibration measurements – a survey. J. Theor. Appl. Mech. 34(2):307-326.

5. Lim TW, Kashangaki TAL (1994) Structural damage detection of space truss structures using best achievable eigenvectors. AIAA Journal 30(9): 2310-2317
6. Farrar C, Jauregui D (1996) Damage detection algorithms applied to experimental and numerical modal data from I-40 bridge. Report No. LA-13074, Los Alamos National Laboratory, Los Alamos, NM
7. Lakshmanan KA, Pines DJ (1997) Detecting crack size and location in composite rotorcraft flexbeams. Proceedings of the SPIE Smart Structures and Materials, 3041: 408-416
8. Lakshmanan KA, Pines DJ (1997) Modelling damage in composite rotorcraft flexbeams using wave mechanics. Smart Materials and Structures 6: 383-392
9. Chang FK (1997) Structural Health Monitoring: Current Status and Perspectives. Edited by F.-K. Chang, Technomic Publishing Co., Inc., Lancaster – Basel
10. Chang FK (1999): Structural Health Monitoring: Current Status and Perspectives. Edited by F.-K. Chang, Technomic Publishing Co., Inc., Lancaster – Basel
11. Chang FK (2001) Structural Health Monitoring: Current Status and Perspectives. Edited by F.-K. Chang, Technomic Publishing Co., Inc., Lancaster – Basel
12. Krawczuk M, Ostachowicz W (2001) Spectral finite element and genetic algorithm for crack detection in cantilever rod. Key Engineering Materials 204-205: 241-250
13. Krawczuk M, Ostachowicz W (2001) Spectral beam finite element and genetic algorithm for crack detection in beams. Proceedings of the 3rd International Workshop on Structural Health Monitoring, Stanford, CA, USA, 889-898
14. Ostachowicz W, Krawczuk M, Cartmell MP, Gilchrist M (2002) Detection of delaminations using spectral finite element. Proceedings of the First European Workshop on Structural Health Monitoring, Paris, France, 344-351
15. Krawczuk M (1994) Dynamika elementów konstrukcyjnych z pęknięciami zmęczeniowymi. Zeszyty Naukowe IMP-PAN, No. 441/1398/94:1-188
16. Krawczuk M, Ostachowicz W (2002) Spectral finite element and genetic algorithm for crack detection in cantilever rod. Key Engineering Materials 204-205: 241-250
17. Krawczuk M, Ostachowicz W (2003) Wave propagation in damaged structures. Proc. 15th International Conference on Computer Methods in Mechanics CMM 2003, 197-199
18. Doyle JF, (1997) Wave Propagation in Structures. Springer–Verlag, New York
19. Krawczuk M, Palacz M, Ostachowicz W (2003) Wave propagation in plate structures for crack detection. Finite Elements in Analysis and Design (in press)
20. Khadem SE, Rezaee M (2000) Introduction of modified comparison functions for vibration analysis of a rectangular cracked plate. J. Sound Vibration 236(2): 245-258
21. Krawczuk M (2002) Application of spectral beam finite element with a crack and iterative search technique for damage detection. Finite Elements in Analysis and Design 38: 537-548
22. Krawczuk M, Palacz M, Ostachowicz W (2003) The dynamic analysis of cracked Timoshenko beam by the spectral element method. J. Sound Vibration 264: 1139-1153

Spatial Filtering with Discrete Array Sensors and Distributed PVDF Films

A. Preumont[1], P. De Man[1], A. François[1], N. Loix[2], and K. Henrioulle[2]

[1] Active Structures Laboratory, Université Libre de Bruxelles, CP165-42, av F.D. Roosevelt 50, B-1050 Brussels, Belgium andre.preumont@ulb.ac.be

[2] Micromega Dynamics sa, Parc Scientifique du Sart Tilman, Rue des Chasseurs Ardennais, B-4031 Angleur, Belgium nloix@Micromega-Dynamics.com

1 Introduction

There are two broad ways to achieve spatial filtering: (i) arrays of discrete sensors and (ii) continuous distributed sensors. Discrete sensor arrays may include accelerometers, strain gages, piezoelectric patches, etc..., while continuous distributed sensors may consist of piezoelectric films or optical fibers (the latter will not be considered in this study). The output of a piezoelectric sensor is a weighted average of the surface strains in the region covered by the electrodes on the film.

Modal filtering was initially proposed as an alternative to state observers to reduce spillover in modal control [1]. Discrete modal filters can be constructed from the orthogonality conditions of the mode shapes; they can also be constructed from modal test data [2, 3, 4]. Discrete piezoelectric array sensors have been considered as modal sensors for beams [5] or as volume velocity sensor for plates [6]. Discrete array sensors, if wired with independent conditioning electronics, are reconfigurable, but they suffer from spatial aliasing when sensing structural modes with wavelengths comparable to the spacing between sensors in the array. Although successful applications of modal control with discrete modal point sensors and actuators have indeed been reported (e.g. [3, 4]), spatial aliasing usually brings strong limitations for applications in structural control, as illustrated later in this paper (a good rule of thumb regarding sensor selection is that the quality of the sensor must in general be guaranteed one decade above the bandwidth of the control system).

When a distributed sensor is located in the far field, that is far from the actuator and from the structural boundaries and singularities, the sensor output can be viewed as the filtered output of a point sensor at the centre of the distributed sensor [7]; the sensor's dynamics are obtained directly from the Fourier transform of the spatial distribution of the sensor. This allows to design distributed sensors with specified low-pass filtering properties featuring high order roll-off without phase lag (with an apparent violation of the Bode's gain-phase relationships). However, in structural control, it is often advantageous to locate the actuator and sensor as close as possible to each other, to produce an interlacing pattern of poles and zeros (such a pattern is strictly achieved for collocated actuator/sensor pairs, but it can still be achieved in low frequency if the actuator and sensor are reasonably close, e.g. [8]. Unfortunately, in this case, the far field condition is violated and the interpretation of the spatial sensor becomes more difficult.

The modal filtering of one-dimensional structures with continuous PVDF films can be traced to [9, 10, 11]. It is achieved by tailoring the width of the electrode (and possibly reversing the polarity). Although the spatial filtering of plates and shells with two-dimensional PVDF films has been suggested [11], it has never been implemented for lack of capability of continuously shaping the piezoelectric properties of the sensor material. A way to turn around this difficulty by a proper electrode design is discussed in this paper.

The use of orthogonal unidirectional PVDF films for spatial filtering of two-dimensional structures has been investigated in [12, 13, 14]; applications of multiple one-dimensional PVDF films to sound radiation sensing has been proposed in [15, 16]. Multiple piezoelectric film sensors have been considered by [17, 18].

This review is by no means exhaustive, and we apologize for many omissions, but it gives an impression of the research effort devoted to this field by the smart structures community. The objective of this paper is to explain the difficulties related to the spatial aliasing that we met in attempting to use discrete piezoelectric array sensors in feedback control loops, and to describe a "porous" electrode design that we developed to solve these problems. The paper is organized as follows: The first part examines the construction of modal filters from an array of discrete piezoelectric patches connected to a linear combiner; the linear combiner coefficients are calculated from a model of the structure, or from experimental FRFs. The tailoring of the open-loop FRF in a frequency band through proper selection of the linear combiner coefficients is also addressed. Next, an experiment is conducted; it confirms the good tailoring at low frequency, but reveals strong departures from the ideal behaviour at higher frequencies, due to spatial aliasing. The spatial aliasing is further analysed numerically on the test configuration with arrays of increasing sizes.

The second part of the paper is devoted to a new type of distributed sensor aimed at overcoming the spatial aliasing of discrete arrays; the distributed sensor is developed as a limit case of a discrete array sensor when the number of elements of the array increases to infinity. Unlike the approach based on tailoring the piezoelectric properties of the sensor, which is impossible to do at this time, this approach can be realized practically through the use of a one-sided porous electrode whose electrode density is adjusted to match the desired effective piezoelectric properties. For a very thin piezoelectric film (when the thickness is small compared to the size of the motif on the electrode), it has been shown by 3-D finite element calculations that the relationship between the effective piezoelectric properties and the fraction of electrode area is almost linear; when the film thickness becomes comparable to the size ot the motif of the electrode, tridimensional (edge) effects appear and the electric field lines are no longer normal to the electrodes, leading to nonlinearities in the relationship between the effective piezoelectric properties and the electrode porosity. The porous electrode concept has been validated on various experiments: (i) modal filter of a cantilever beam [19, 20] (ii) volume velocity sensor of a simply supported plate [20, 21], and a transparent implementation of the volume velocity has been realized [21]; finally, a comparison of the electrode design for orthotropic (uni-axial) and isotropic (copolymer) PVDF films is conducted.

This paper synthesizes the results presented previously in [19, 20, 21, 22].

2 Modal Filtering with an Array Sensor

2.1 From a Known Model of the Structure

We first assume that we have an accurate linear model of a known structure. The modal expansion of the FRF of the sensor array reads

$$Y_k(\omega) = \sum_{i=1}^{m} \frac{c_{ki} b_i}{\mu_i(\omega_i^2 - \omega^2 + 2j\xi_i\omega_i\omega)} \qquad k = 1, ..., n \tag{1}$$

where b_i is the modal input gain (at the actuator) and c_{ki} is the modal output gain of sensor k in the array.

If the n sensors in the array are connected to a linear combiner with gain α_k for sensor k (Fig. 1), the output of the linear combiner is $y = \sum_{k=1}^{n} \alpha_k y_k$ and the global frequency response is

$$G(\omega) = \sum_{k=1}^{n} \alpha_k Y_k(\omega) = \sum_{i=1}^{m} \frac{\{\sum_{k=1}^{n} \alpha_k c_{ki}\} b_i}{\mu_i(\omega_i^2 - \omega^2 + 2j\xi_i\omega_i\omega)} \tag{2}$$

A modal filter which isolates mode l can be constructed by selecting the weighing coefficients α_k of the linear combiner in such a way that

$$\sum_{k=1}^{n} \alpha_k c_{ki}(\omega) = \delta_{li} \tag{3}$$

or

$$\boldsymbol{\alpha}^T C = \boldsymbol{e}_l^T \tag{4}$$

where $\boldsymbol{\alpha} = (\alpha_1 \dots \alpha_n)^T$ is the vector of the linear combiner coefficients, $C = c_{ki}$ is the matrix of modal output gain (column i is the sensor array output when the structure vibrates according to mode i) and $\boldsymbol{e}_l = (0\ 0 \dots 1 \dots 0)^T$

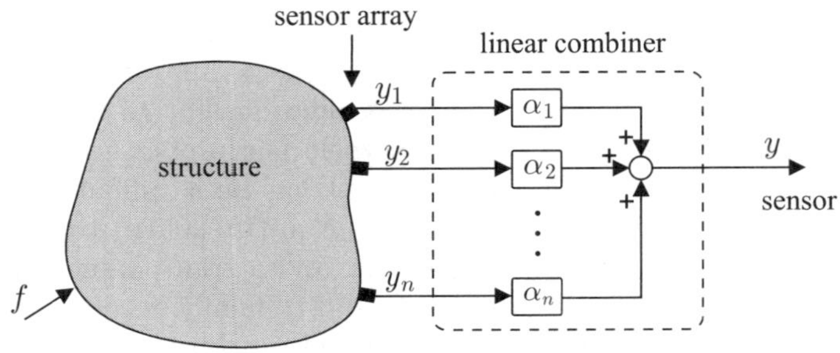

Fig. 1. Representation of the modal filter using n discrete sensors

is the vector with all entries equal to 0 except entry l which is equal to 1. Assuming that matrix C is known accurately, the modal filter coefficients $\boldsymbol{\alpha}_l$ for mode l can be found by solving the rectangular system of equations

$$C^T \boldsymbol{\alpha}_l = \boldsymbol{e}_l \tag{5}$$

The number of columns of C^T, n, is equal to the number of sensors in the array and the number of lines, m, is equal to the number of modes included in the truncated modal expansion (1). The solution of Eq. (5) is discussed next section.

If the set of weighing coefficients $\boldsymbol{\alpha}_l$ satisfy Eq. (4), the open-loop FRF of the system, Eq. (2), becomes

$$G(\omega) = \frac{b_l}{\mu_l(\omega_l^2 - \omega^2 + 2j\xi_l\omega_l\omega)} \tag{6}$$

Its maximum value is obtained at $\omega = \omega_l$; it is equal to [23]

$$\left\|G(\omega)\right\|_\infty = \frac{b_l}{2\mu_l\xi_l\omega_l^2} \tag{7}$$

2.2 Modal Filter Coeficicients $\boldsymbol{\alpha}_l$

The solution of equation (5) requires some care because of the rank deficiency of C connected to the spatial aliasing. In fact, if the column of C (that is the modal contributions of the array sensor) are independent, the rank r of C follows the continuous line in Fig. 2 : $r = m$ as long as $m \le n$ and $r = n$ thereafter. This is the case, for example, for a simply supported beam with a regular array sensor, or for a simply supported square plate with a square array sensor. However, it is easy to check that for a *rectangular* plate with a uniform array sensor, the columns of C are not independent, resulting in a rank deficiency as indicated in dotted line in Fig. 2. The spatial aliasing is illustrated in Fig. 9 which shows two columns of C, corresponding respectively to mode (1,1) (mode # 1) and mode (1,15) (mode # 69). Similarly, mode (1,9)

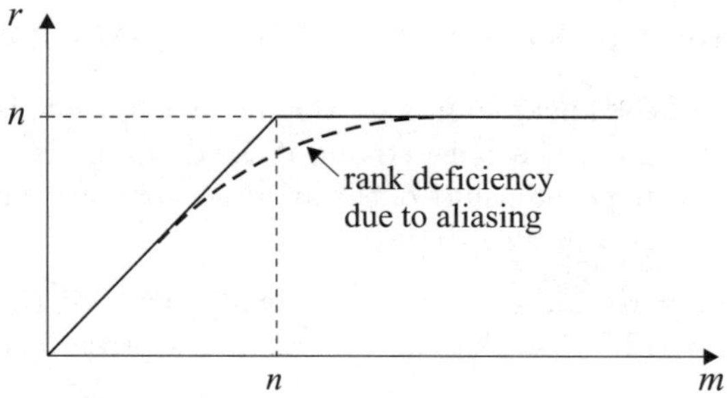

Fig. 2. Rank r of $C(m,n)$ as a function of the number of modes m in the modal expansion (1)

(# 24) will be aliased into mode (1,7) (# 14). Spatial aliasing will be discussed further later in this paper.

The difficulty associated with the rank deficiency of C can be overcome by using a singular value decomposition of C^T [24].

$$C^T = U_1 \Sigma U_2^H \tag{8}$$

where U_1 and U_2 are unitary matrices containing the eigenvectors of $C^T C$ and CC^T respectively, and Σ is the rectangular matrix of dimension (m,n) with the singular values σ_i on the diagonal (the subscript H is used to indicate the Hermitian). If $\mathbf{u}_i$ are the column vectors of U_1 and $\mathbf{v}_i$ the column vectors of U_2 Eq. (8) can be written

$$C^T = \sum_{i=1}^{r} \sigma_i \mathbf{u}_i \mathbf{v}_i^H \tag{9}$$

and the solution of Eq. (5) reads

$$\boldsymbol{\alpha}_l = (C^T)^+ \boldsymbol{e}_l \tag{10}$$

where the pseudo-inverse reads

$$(C^T)^+ = \sum_{i=1}^{r} \frac{1}{\sigma_i} \mathbf{v}_i \mathbf{u}_i^H \tag{11}$$

This equation shows that the lowest singular values tend to dominate the pseudo-inverse, which is the origin of the problem in solving Eq. (5). In practice, however, the columns of C are not strictly proportional and its exact rank is not always easy to determine. The problem can be solved by truncating all the singular values below some tolerance value. In general, the solution is easier to achieve when the modal truncation is such that $m < n$.

2.3 Tailoring the Open-Loop FRF with the Array Sensor

Let $\boldsymbol{\alpha}_l$ be the set of the linear combiner weighing coefficients leading to modal filters. If the modes are well separated and if the damping is low, it is possible to find a set of weighing coefficients of the array sensor such that the open-loop FRF features the following properties:

1. The poles and zeros alternate near the imaginary axis (interlacing).
2. The peaks of the FRF at the resonances have a prescribed amplitudes A_i (Fig. 3).

The weighing coefficients can be obtained as follows: If the damping is low and the modes are well separated, the magnitude at a resonance peak is dominated by the contribution of the corresponding mode. Equation (7) gives the maximum amplitude achieved with the linear combiner coefficients $\boldsymbol{\alpha}_l$. It follows that if the coefficients are taken as $w_l \boldsymbol{\alpha}_l$ with

$$w_l = \frac{2\mu_l \xi_l \omega_l^2}{b_l} A_l \tag{12}$$

the maximum amplitude of the response will be exactly A_l ($A_l > 0$). Since, for low damping and well separated modes, the amplitude at resonance is dominated by the contribution of the resonant mode, it follows that the set of weighing coefficients

$$\boldsymbol{\alpha} = \sum_{l=1}^{M} w_l \boldsymbol{\alpha}_l \tag{13}$$

will lead to an open loop FRF with a set of fixed amplitudes A_l at the resonances of the M selected modes [23]. Besides, if one substitutes Eq. (12) and

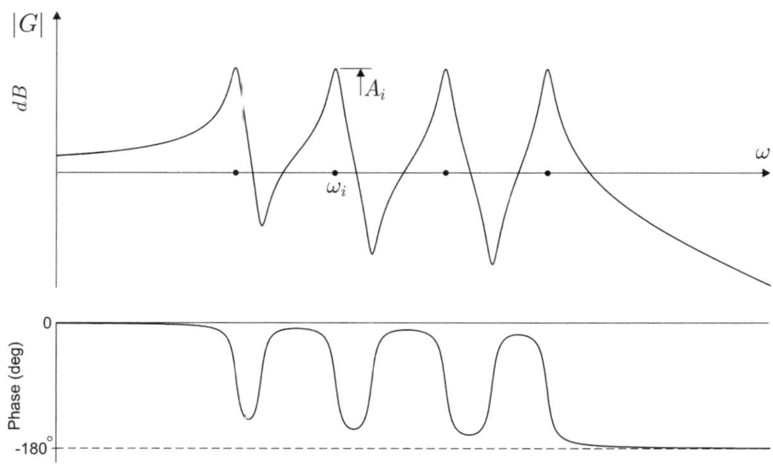

Fig. 3. Construction of a FRF with resonance peaks of prescribed amplitudes A_i and alternating poles and zeros

(13) in Eq. (2), the FRF reads

$$G(\omega) = \sum_{i=1}^{M} \frac{2\xi_i \omega_i^2 A_i}{\omega_i^2 - \omega^2 + 2j\xi_i\omega_i\omega} \tag{14}$$

Since all the residues are positive, alternating poles and zeros are guaranteed [25].

2.4 From Experimental Data

We now address the following problem: The individual FRF $Y_k(\omega)$ of the n sensors in the array have been measured and the natural frequency ω_i and the modal damping ξ_i of one mode have been determined (with a modal analysis software). What are the weighing coefficients α_k of the linear combiner leading to a modal filter within some bandwidth $[\omega_a, \omega_b]$ and with unit amplitude at resonance (Fig. 4)

$$G_i(\omega) = \frac{2\xi_i \omega_i^2}{\omega_i^2 - \omega^2 + 2j\xi_i\omega_i\omega} \tag{15}$$

The output FRF of the linear combiner reads

$$G(\omega) = \sum_{k=1}^{n} \alpha_k Y_k(\omega) \tag{16}$$

and the weighing coefficients α_k satisfy

$$\sum_{k=1}^{n} \alpha_k Y_k(\omega) = G_i(\omega) \tag{17}$$

If this equation is written at a set of p discrete frequencies ν_i $(p > n)$ regularly distributed over the frequency band $[\omega_a, \omega_b]$ the function equality (17) can be transformed into a redundant system of linear equations

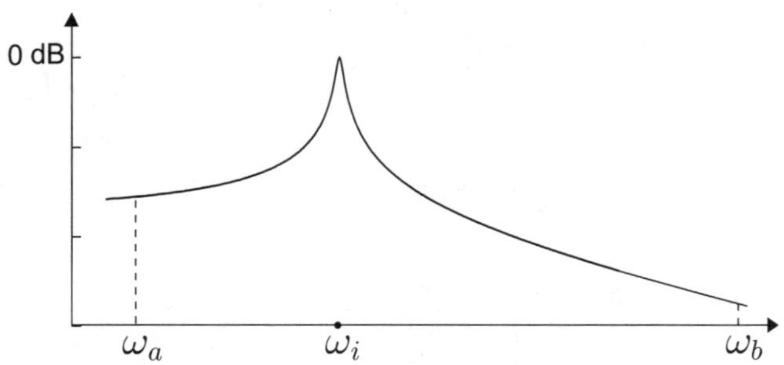

Fig. 4. Perfect modal filter within the bandwidth $[\omega_a, \omega_b]$ and with unit amplitude at ω_i

$$\begin{pmatrix} Y_1(\nu_1) & \dots & Y_n(\nu_1) \\ Y_1(\nu_2) & \dots & Y_n(\nu_2) \\ & \dots & \\ Y_1(\nu_p) & \dots & Y_n(\nu_p) \end{pmatrix} \begin{pmatrix} \alpha_1 \\ \alpha_2 \\ \vdots \\ \alpha_n \end{pmatrix} = \begin{pmatrix} G_i(\nu_1) \\ G_i(\nu_2) \\ \vdots \\ G_i(\nu_p) \end{pmatrix} \tag{18}$$

or, in matrix form,

$$Y\boldsymbol{\alpha} = \boldsymbol{G}_i \tag{19}$$

where the rectangular matrix Y, of dimension (p, n), and the vector $\boldsymbol{G}_i$ are complex quantities and the vector $\boldsymbol{\alpha}$ of the linear combiner coefficients is real.

The solution of this redundant system of linear equations was addressed in [6]; the pseudo-inverse in the mean-square sense, $Y^+ = (Y^T Y)^{-1} Y^T$ produces highly irregular coefficients. As in the previous section, the difficulty can be solved using a singular value decomposition of Y and truncating the contribution of the least significant singular values.

Numerical simulations have shown that in a system without noise, the rank of the system is equal to the number of modes which respond significantly in the frequency band of interest (assuming this number is smaller that the number n of sensors in the array). When dealing with actual experimental data, the gap in magnitude between significant and insignificant singular values disappears and some trial and error is needed to select the appropriate number of singular values in the truncated expansion.

3 Experiment

Figure 5 shows a view of the experimental set-up that consists of a 4 mm thick glass plate (0.54 m × 1.24 m) mounted in a standard window frame that is fixed on a concrete box. The sensor array consists of 4 × 8 piezoceramic (PZT) patches (13.75 mm × 25 mm) glued on the plate according to a regular mesh. The set-up and the linear combiner hardware are described in [6] where they were used in a volume velocity sensor. The resonance frequencies and structural damping of the plate are given Table 1.

Figure 6 shows the weighing coefficients α_i leading to modal filters for the first four modes of the plate, obtained using Eq. (15) and (19). One can tailor a FRF similar to Fig. 3 with alternating poles and zeros and unit amplitudes at the resonances. Figure 7 shows the FRF measured between the

Table 1. Natural frequencies and structural damping of the plate.

Mode	Frequency	Damping
(1,1)	42.5 Hz	2.31 %
(1,2)	55.9 Hz	1.08 %
(1,3)	87.1 Hz	1.24 %
(1,4)	118.7 Hz	2.42 %

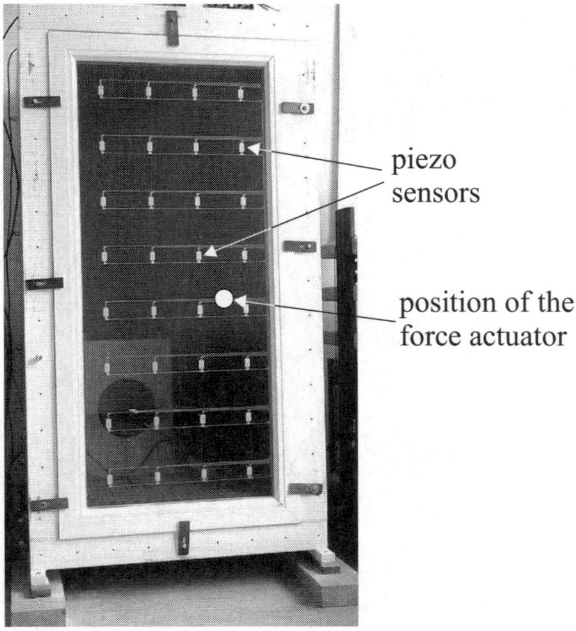

Fig. 5. Experimental set-up: glass plate covered with an array of 4×8 piezoelectric patches

actuator and the sensor output with weighing coefficient given by Eq. (13). We observe an excellent agreement at low frequency, but the measured FRF departs substantially from the expected one at higher frequency.

4 Spatial Aliasing

As illustrated in the previous section, reconfigurable sensor arrays work very well as modal filters in a limited frequency band. However, they are subject to spatial aliasing which degrades their behaviour beyond this frequency band by reducing the sensor roll-off. This is again illustrated with the analytical example of Fig. 8. The linear combiner coefficients of the 4×8 PZT array sensor are selected to isolate the first mode. Figure 8 shows the FRF between four point force actuators acting in phase (same current applied to all) and located as indicated in Fig. 8.(a), and the linear combiner output. When the sensor is used in an active vibration control system, this reduces the roll-off of the open-loop transfer function, which imposes strong limitations on the bandwidth of the control system (a good sensor should have a bandwidth at least one decade larger than that of the control system).

The spatial aliasing is the counterpart of the better known time aliasing: when the wave number (k) of one mode exceeds the number (n) of sensors

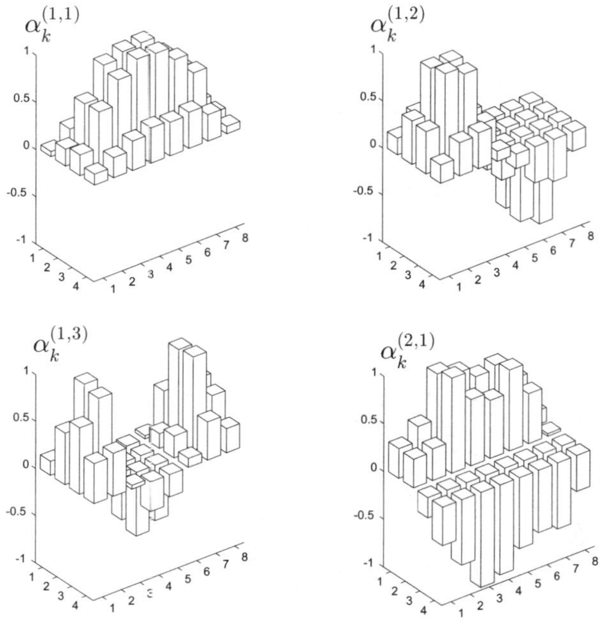

Fig. 6. Weighing coefficients α_i (normalized) leading to modal filters for the first four modes of the plate

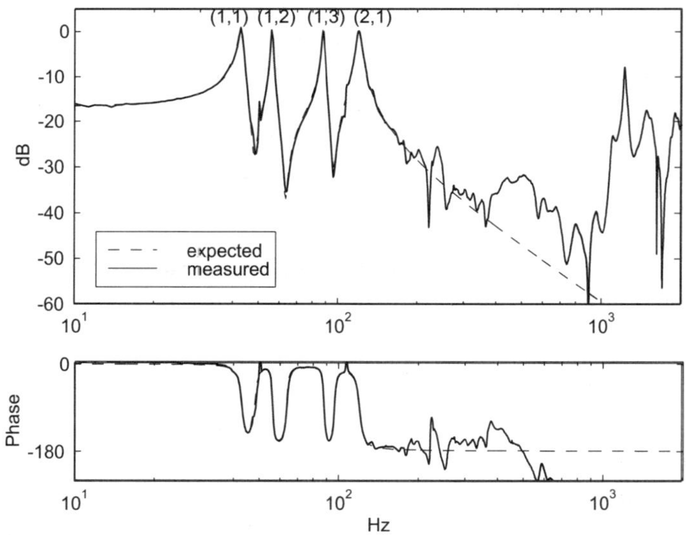

Fig. 7. FRF with prescribed unit amplitude at the resonant peaks, obtained with weighing coefficients $\alpha = \alpha_{(1,1)} + \alpha_{(1,2)} + \alpha_{(1,3)} + \alpha_{(2,1)}$. Comparison between expected and experimental results

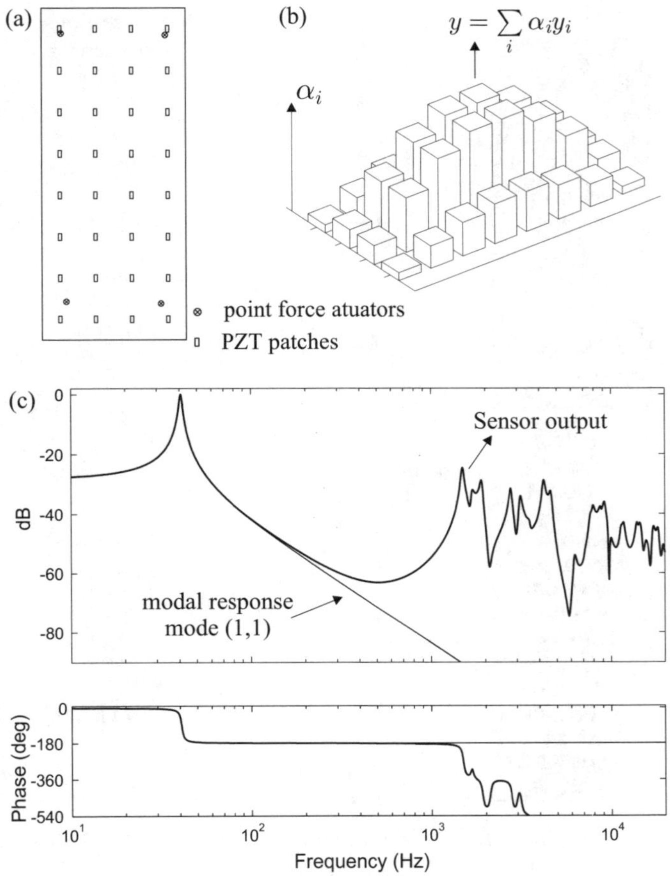

Fig. 8. (a) Geometry of the 4 × 8 array sensor, (b) Weighing coefficients α_i of the linear combiner, (c) Comparison of the FRF between the actuators and the modal response of mode (1,1) and the sensor output (numerical simulation)

regularly spaced in that direction, the sensor output appears as generated by a mode with a lower wave number $(2n-k)$. It is illustrated in Fig. 9 for the plate considered previously. The left part of Fig. 9 shows the mode shapes (1,1) and (1,15); the diagrams on the right show the electric charges Q_i generated by these modes on the PZT patches (numerical simulation). The electric charges generated by mode (1,15) at 1494 Hz have the same shape as those generated by mode (1,1). As a result, the orthogonality relationship (4) between the linear combiner coefficients α_i and the sensor modal gains cannot be enforced and the modal sensor designed for mode (1,1) responds to mode (1,15) as well. The limit frequency of the modal filter is given by the natural frequency of the mode with wavenumber equal to the size of the sensor array.

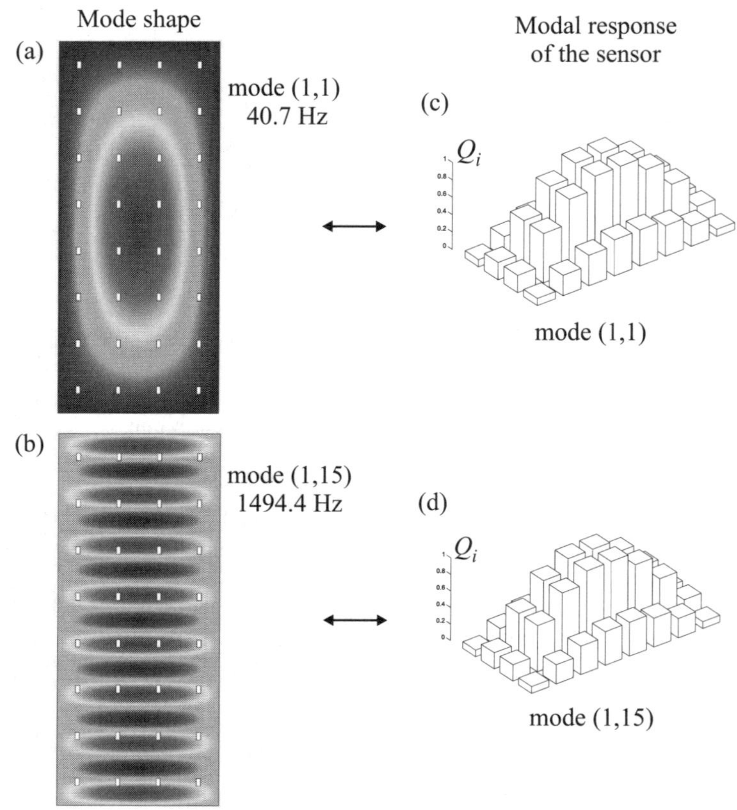

Fig. 9. Spatial aliasing: (a), (b) mode shapes (1,1) and (1,15); (c), (d) electric charges Q_i generated by mode (1,1) and mode (1,15)

Figure 10 shows a numerical simulation of the influence of the sensor array on the open-loop FRF of a modal filter for mode (1,1). Because of the high modal density, wide-band modal filtering may require a sensor array of large size which may not be practical because of the independent conditioning electronics, but this is unescapable with discrete array sensors.

5 Distributed Sensor

As we have seen in the previous section, the need for distributed sensors arises from the necessity to reduce the spatial aliasing. Although the theory of modal distributed piezoelectric sensors has been known for some time [10, 11], they have never been implemented in practice to two-dimensional structure, because the theory is based on tailoring the piezoelectric coefficients, which cannot be done in practice at this time. In this section, a *practical* way of designing a distributed modal filter is proposed.

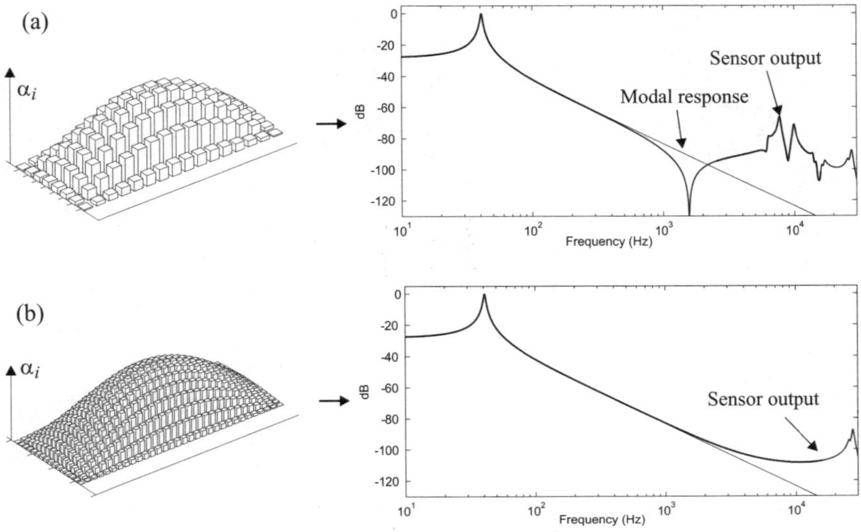

Fig. 10. Effect of the size of the array on the open-loop FRF (a) 8×16, (b) 16×32

The theory of modal filtering with distributed sensors can be addressed in two different ways: (i) based on the orthogonality relationships for distributed structures or (ii) as a limit case of a discrete array filter when the number of elements in the array increases to infinity. The first approach was addressed by Lee and Moon; we will consider the second one which is more intuitive in view of the previous discussion on spatial aliasing.

As illustrated in Fig. 10, the bandwidth of an array sensor can be increased by increasing the number of sensing elements in the array. In the example of Fig. 10.b, the sensor bandwidth is close to 5000 Hz with a sensor array containing 16×32 elements. However, the number of independent conditioning electronic units becomes rapidly prohibitive.

If the weighing coefficients α_i are known before hand and if we give up the programmability of the linear combiner, the coefficients α_i can be embedded in the individual sensing elements. This is illustrated in Fig. 11; the area of the individual piezoelectric patches (more precisely the electrode area) is taken proportional to the corresponding coefficient α_i in the linear combiner. The various electrodes of the sensor array are connected together and the total electric charge generated on the array is proportional to the output of the linear combiner. This configuration can be easily manufactured with a segmented electrode etched on a continuous PVDF layer using classical lithography techniques; it requires only a single amplifier, but the weighing coefficients cannot be changed on-line as in the linear combiner. The modal filtering property of this concept is illustrated in Fig. 11; spatial aliasing still occurs beyond 2000 Hz and it can be pushed even further by increasing the number of segments in the electrode.

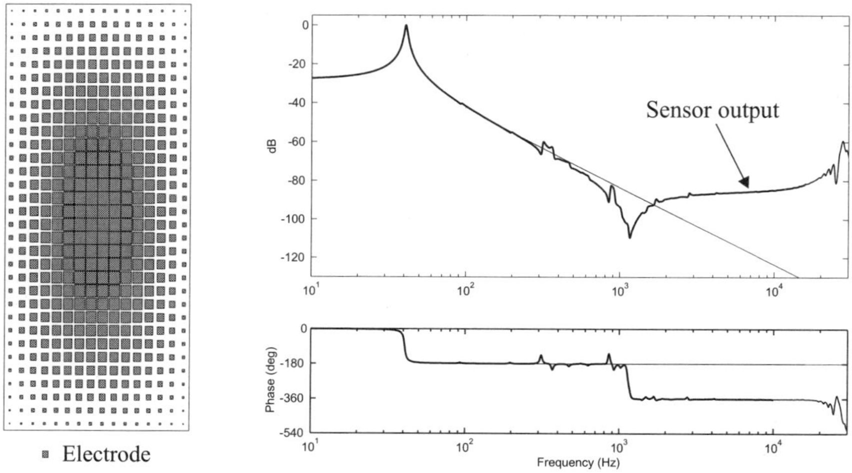

Fig. 11. Variable size array (the 16 × 32 patches are interconnected)

6 Porous Electrode Concept

The distributed sensor can be viewed as the limit discrete array sensor where the number of sensing elements increases and the electrode density is such that the local production of electric charges matches the desired local effective weighing coefficient. A practical way to achieve this is to use a "porous" electrode as shown in Fig. 12; the maximum sensitivity is obtained where the electrode is continuous (in the center in Fig 12.a) and the local sensitivity is decreased continuously by introducing some porosity in the continuous electrode by means of a honeycomb design (Fig. 12.b). The local electrode density is selected in such a way that the local production of electric charges matches the desired local weighing coefficient $\alpha(x, y)$.

Consider a piezoelectric film polarized in the direction normal to its plane and covered with electrodes Ω as in Fig. 13. According to the two-dimensional theory of piezoelectric films, if a sample is subject to a plane strain field aligned on the orthotropy axes of the material, and if the electrodes are connected to a charge amplifier which cancels the electric field across the piezoelectric film, the electric charge produced on the electrodes is

$$Q = \int_{\Omega} (e_{31}S_1 + e_{32}S_2)d\Omega \tag{20}$$

where S_1 and S_2 are the strain components along the orthotropy axes in the mid-plane of the film, and e_{31} and e_{32} are the piezoelectric constants of the material. The integral extends over the area of the electrode, or more precisely, the area Ω over which the two electrodes overlap, where the electrical field is zero. According to Eq. (20) changing the local electrode density ($d\Omega$) is

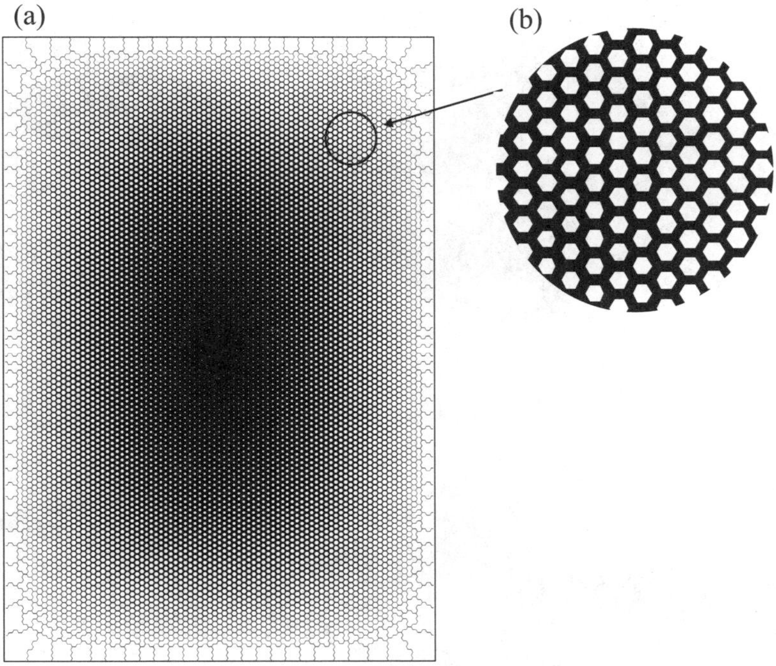

Fig. 12. (a) Porous electrode, (b) detail of the motif with variable porosity

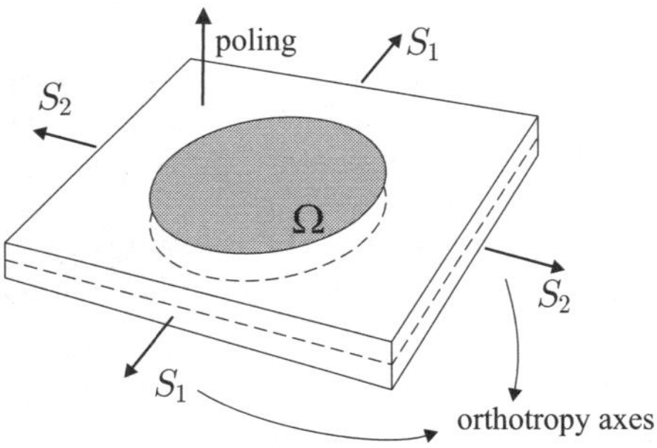

Fig. 13. Sample of piezoelectric film polarized in the direction normal to its plane

equivalent to changing the piezoelectric coefficients e_{31} and e_{32} in the same ratio. Since Ω refers to the area where the two electrodes overlap, the motif may be etched on the electrodes on both sides of the piezo film (Fig. 14.a) or only on one side, with a continuous electrode on the other side (Fig. 14.b), which seems technologically simpler, because the two electrodes do not have to be aligned precisely on top of each other. We will consider only the one-sided design in what follows.

Equation (20) assumes that the size of the electrode is much larger than its thickness. However, when the motif of the electrode becomes small, tridimen-

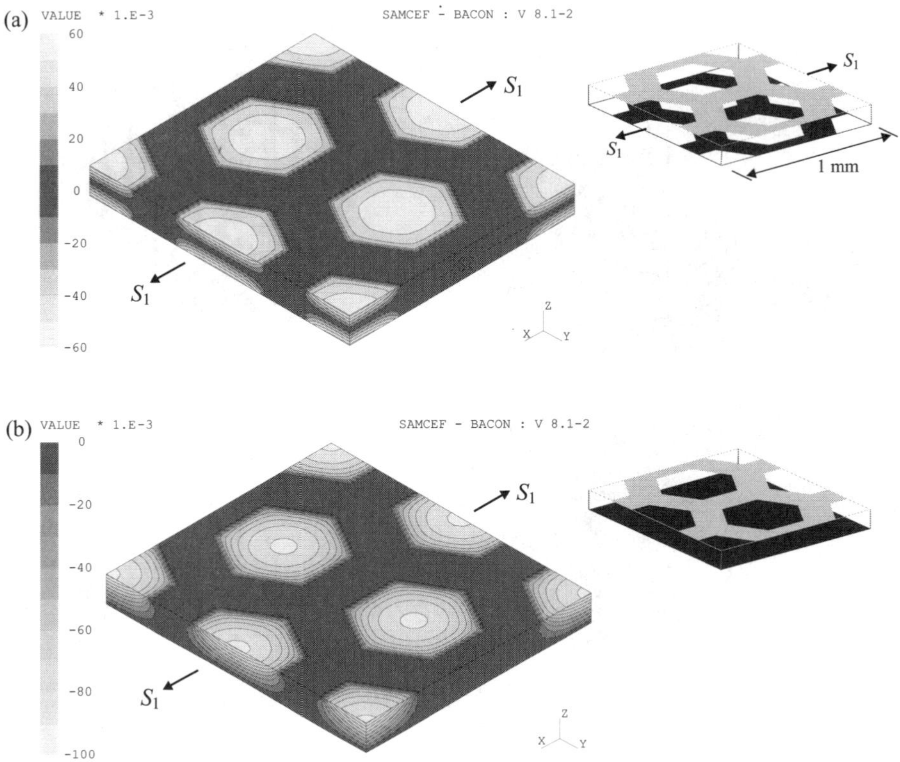

Fig. 14. Tridimensional finite element analysis. Isopotential surfaces (a) Two-sided electrode (fraction of electrode area = 50%) (b) One-sided electrode

sional (edge) effects start to appear and the relationship between the porosity and the equivalent piezoelectric property is no longer linear [19].

7 Design of the Porous Electrode

The electrode design starts with a discrete array of large size which is used as a discretization of the distributed sensor. The linear combiner coefficients achieving the required spatial filtering properties are calculated according to the procedure described earlier; this is illustrated in Fig. 15.a.

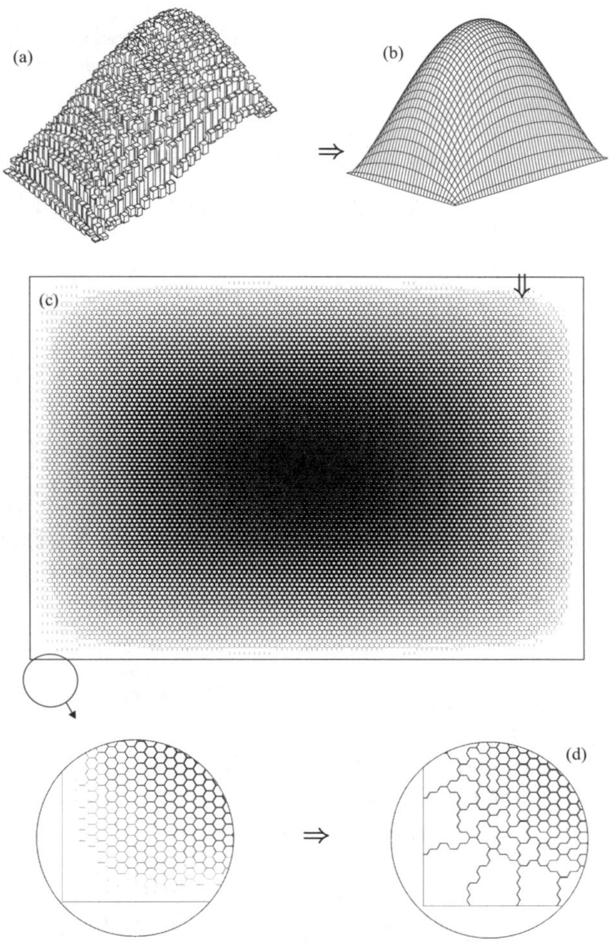

Fig. 15. Various steps of the design of the porous electrode: (a) computation of the weighing coefficients of the linear combiner, (b) smoothing of the coefficients to obtain the polarization profile, (c) transformation of the polarization profile into the fraction of electrode area leading to the honeycomb electrode profile, (d) redesign of the edge of the electrode to take care of the manufacturing tolerances and electrical connections

The polarization profile of the distributed spatial filter is then obtained by interpolating and smoothing the linear combiner coefficient diagram as in Fig. 15.b; there is no obstacle in handling negative coefficients, by changing the polarity of the PVDF film. Next, the polarization profile can be converted into porosity profile of the honeycomb electrode (Fig. 15.c) using the relationship between the effective piezoelectric properties and the fraction of electrode area

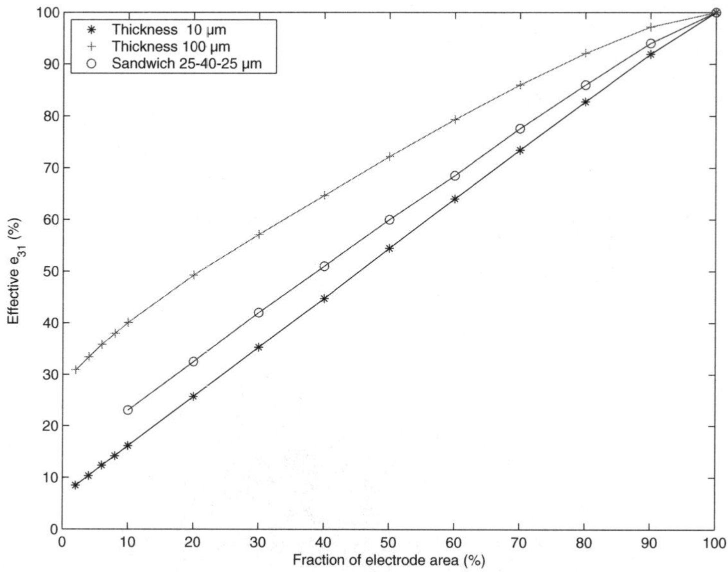

Fig. 16. Effective piezoelectric coefficient versus fraction of electrode area for an isotropic PVDF film of various thickness provided with a one-sided honeycomb electrode

(Fig. 16), obtained by homogenization from 3-D finite element calculations of the piezo film with appropriate electrical boundary conditions (potential difference $V = 0$ between the electrodes in this case).

If the thickness of the film is small compared to the size of the motif, this relation can be assumed linear. The final step consists of adapting the edge of the electrode to take into account the technological constraints of the lithography process, in particular to keep the minimum thickness of the motif to a value compatible with manufacturing tolerances and to allow the electrical connection with the conditioning electronics (Fig. 15.d).

The following sections report on the test program aiming at validating the porous electrode concept.

8 Modal Filter of a Cantilever Beam

Figure 17 shows the experimental set-up. Two cantilever beams made of glass of 240 mm × 27 mm × 1.83 mm are equipped on one side with a PZT actuator and on the opposite side with an isotropic PVDF (copolymer) sensor with a shaped electrode. In one case, the electrode profile matches the modal filter of the beam theory [10]; in the other case, the same axial variation of the piezoelectric properties is achieved with the porous design described in the previous section. However, to guarantee the continuity of the honeycomb electrode during the manufacturing process (accurate etching of the silver electrode on PVDF turned out to be difficult), the width of the electrode is constrained to be larger than 0.5 mm; this explains why the electrode design is tapered near the free end of the sample.

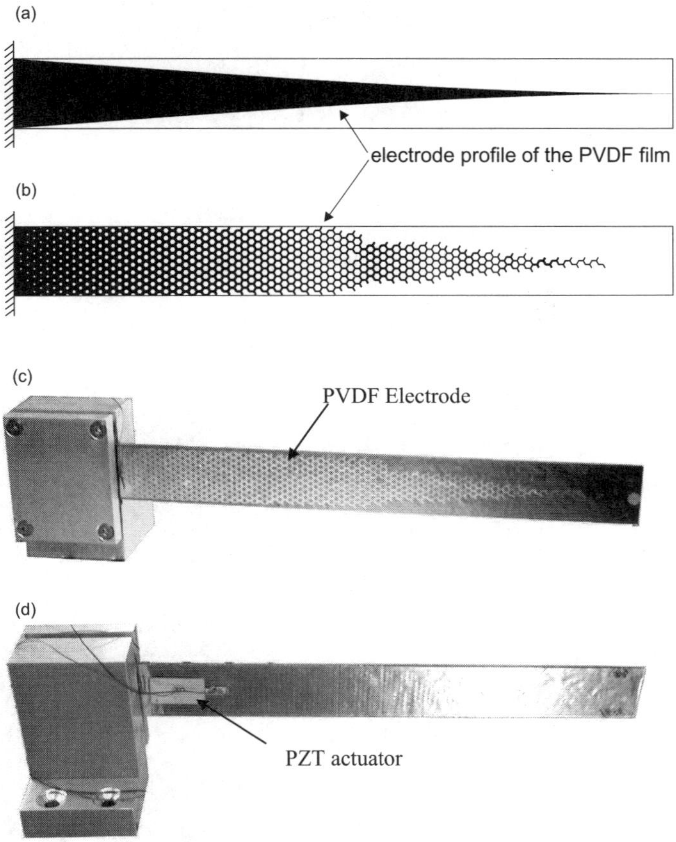

Fig. 17. Experimental set-up for the honeycomb electrode validation

Figure 18 compares the FRF between the PZT actuator and the PVDF sensor for the two electrode designs. The two curves are remarkably close, although significantly different from the ideal modal filter; they also agree well with finite element calculations (Mindlin shell [26]) for the continuous electrode sensor, Fig. 19. This experiment seems to indicate that, when equipped with an isotropic PVDF sensor ($e_{31} = e_{32}$) the beam behaves in a significantly different manner from that predicted by the beam theory.

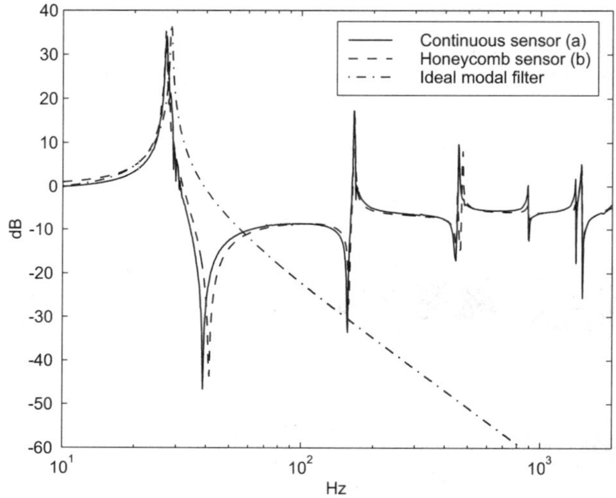

Fig. 18. Comparison between the measured FRF of the continuous and honeycomb sensors with the FRF of the analytical beam model

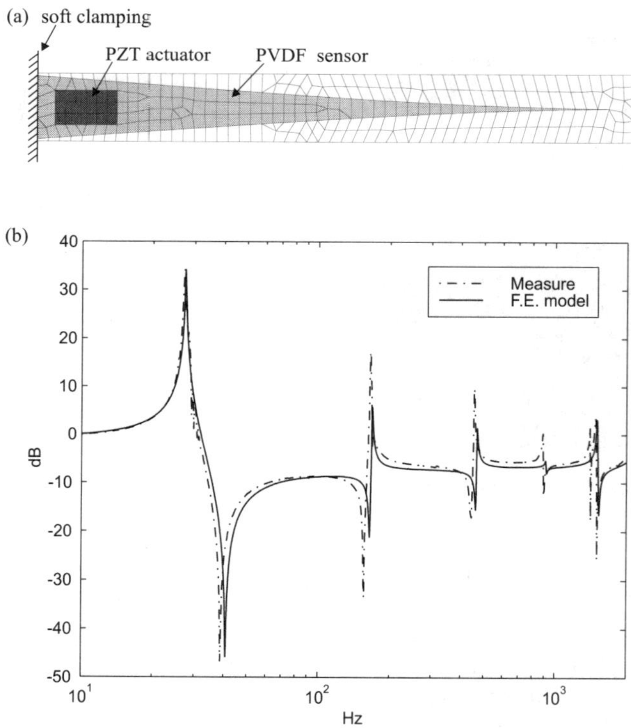

Fig. 19. (a) Finite element model of the beam with its piezoelectric modal sensor, (b) and comparison between the finite element and the experimental results (continuous sensor)

9 Volume Displacement Sensor for a Baffled Plate

This section considers the design and construction of a volume displacement sensor of a baffled simply supported plate. The volume velocity of a vibrating plate is defined as

$$\dot{V} = \int_S \dot{w}\, dS \tag{21}$$

where w is the transverse displacement of the plate and the integral extends over the entire plate area. It is a very important quantity in vibroacoustics, because it closely approximates the first radiation mode which is the dominant contributor to the sound power radiation at low frequency [27]. It is very tempting to try to develop a set of structural sensors measuring directly the volume velocity, or equivalently the volume displacement (by using a charge amplifier instead of a current amplifier). Elliott and co-workers [16] developed the so-called *Quadratically Weighted Strain Integrator Sensor (QWSIS)* which consists of slicing the electrode into narrow strips with parabolic shape. The sensor is based on the assumption that the volume displacement of each strip can be based on the beam theory. This sensor is biased because strictly unidirectional PVDF does not exist (e_{32} is always at least 10% of e_{31}) and it is prone to spatial aliasing due to the finite number of strips. Our research group pursued an alternative strategy based on discrete array sensors which were simpler and reconfigurable [6]; it worked well at low frequency (below 250Hz) where the sensor is really useful, but the sensor output was much degraded at higher frequency because of aliasing, which made it difficult to integrate in a control system [28]. This failure to build a discrete array volume displacement sensor with adequate high frequency behaviour was the origin of the present work.

In this study, we have designed, built and tested a volume displacement sensor with the porous electrode technology. The test article is shown in Fig. 20; it consists of a 180 mm x 260 mm x 1.6 mm glass epoxy (PCB) board with a $37\mu m$ thick copper electrode. The electrode is full in the center of the plate and a honeycomb motif with variable width appears when one moves towards the edges; the honeycomb motif is etched with standard PCB technology in the copper electrode and a $40\mu m$ thick PVDF copolymer film with a one-sided continuous electrode is glued on top of it to form the distributed sensor with variable effective piezoelectric properties. Accurate etching of a PCB copper electrode is technologically much simpler than the silver electrode of a PVDF film as used in the validation test on the beam discussed earlier, and it is possible to achieve a better resolution here (better than $100\,\mu m$, instead of $500\,\mu m$ with silver electrodes on PVDF).

The design of the electrode was done with an analytical model of the simply supported plate, using analytical mode shapes and neglecting the membrane strains in the strain-induced piezoelectric charges [28]; if necessary (especially for more complicated boundary conditions), finite element simulations could

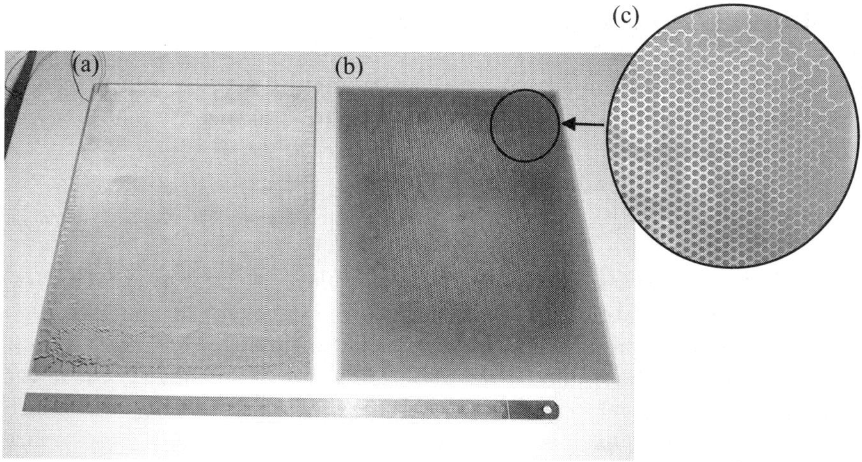

Fig. 20. (a) Top view of the PVDF film mounted on the plate, (b) PCB plate without PVDF film, showing the honeycomb etched in the copper electrode, (c) detail of the honeycomb electrode with variable porosity

also be used at this stage [26], but the analytical model turned out to be sufficient in this case. The procedure for determining the linear combiner coefficients follows Fig. 15.

Figure 21 shows the experimental set-up to assess the quality of the sensor; the plate is first placed in an aluminium frame with an elastomer joint, to approximate simply supported boundary conditions; the aluminium frame in mounted in the opening of a concrete acoustic chamber with a loudspeaker inside. The volume displacement is measured with a scanner laser vibrometer (the mesh used in the laser measurement is also shown in Fig. 21).

Figure 22 compares the FRF between the current applied to the loudspeaker and the volume displacement sensed with (i) the PVDF sensor with porous electrode (referred to as piezo sensor) and (ii) the laser vibrometer (referred to as volume displacement); the curves are very close to each other; the phase diagrams are also close to each other. Figure 23 shows the same comparison for a point force applied with a voice coil acting near a corner, at 35 mm from the edge of the plate. The agreement between the curves is still good, although larger differences are observed; part of these are due to the fact that the coil adds a discrete mass of 1 *gr* near the corner of the plate, altering its mode shapes (in particular some anti-symmetric modes which did not contribute to the volume displacement for the original plate). Thus, the porous sensor which was designed for producing the volume displacement of a symmetric plate is no longer totally appropriate. Overall, the quality of the volume displacement sensor looks quite good.

Fig. 21. Experimental set-up used to test the volume displacement sensor and (upper left) measurement mesh used for the laser vibrometer

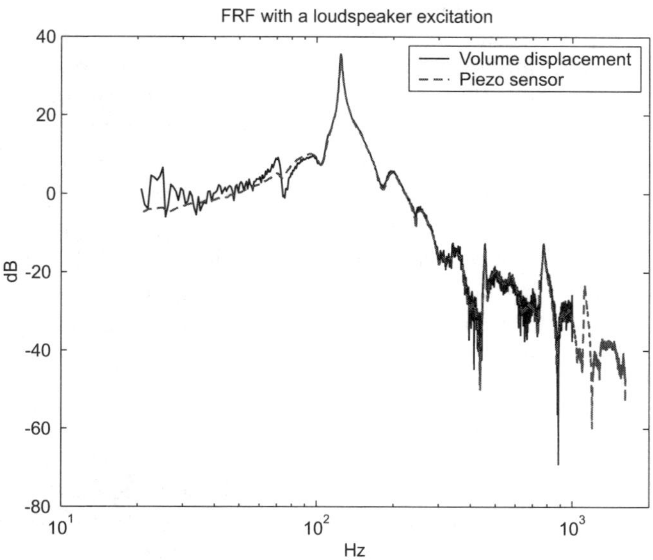

Fig. 22. Glass-epoxy plate. FRF between the current applied to the loudspeaker and the volume displacement; comparison between the porous piezoelectric sensor (dotted line) and the laser vibrometer (full line)

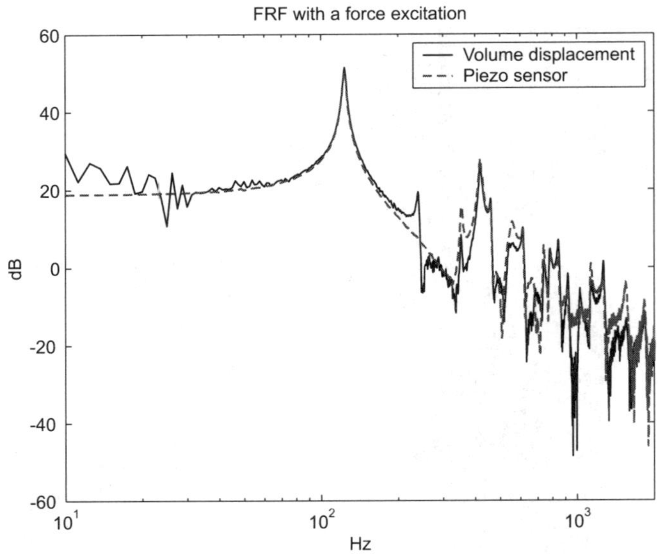

Fig. 23. Glass-epoxy plate. FRF between a point force excitation near a corner of the plate and the volume displacement; comparison between the porous piezoelectric sensor (dotted line) and the laser vibrometer (full line)

9.1 Transparent Implementation

Transparent sound radiation sensors could open the way to new applications such as active windows. Figure 24 shows a transparent implementation of the volume displacement sensor. It consists of a $40\,\mu m$ PVDF film glued on a glass plate with $25\,\mu m$ connecting layers. The transparent electrodes are made of ITO (Indium Tin Oxide). A laser lithography process is used to shape the electrode. The test article has the same size as the glass-epoxy plate tested above and it has been tested with the same set-up of Fig. 21. Figure 25 shows a comparison of the FRF between the loudspeaker input and the sensor output obtained respectively with the transparent sensor and the laser vibrometer. Once again, the agreement between the FRFs is excellent.

9.2 Orthotropic PVDF Sensor

The results of Fig. 25 have been obtained with isotropic copolymer exhibiting isotropic piezoelectric properties ($d_{31} = d_{32} = 2.5pC/N$). Uni-axial PVDF has several advantages over copolymer: it is more sensitive, it is a lot cheaper and it can be more easily mass produced. An alternative sensor has been designed and manufactured with uni-axial PVDF with the following properties: $d_{31} = 22.5\ pC/N$, $d_{32} = 2.5\ pC/N$. The design follows the same lines as before, except that the orthotropy of the sensor must be included in the model,

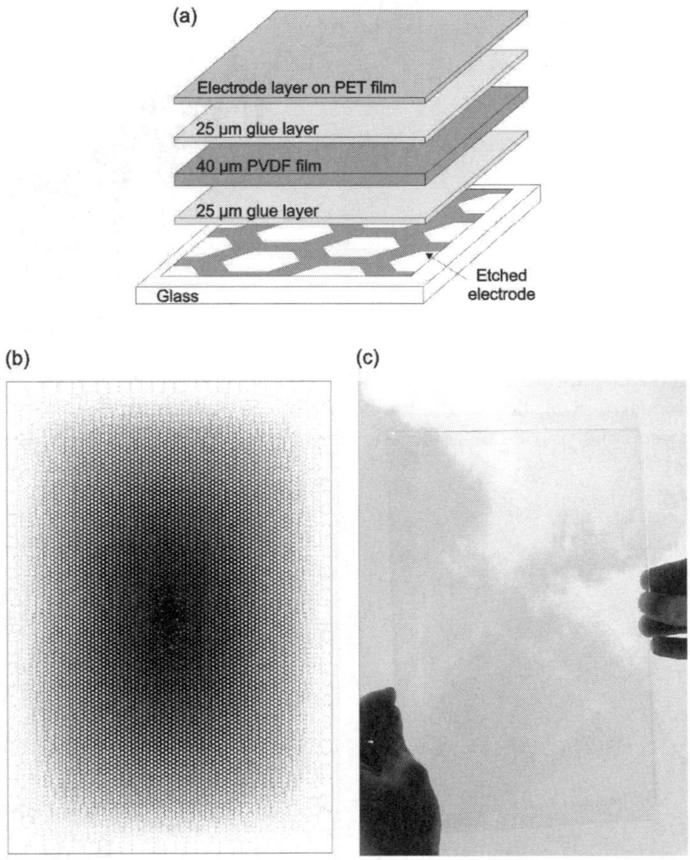

Fig. 24. Transparent PVDF sensor tailored for measuring the volume displacement (a) detail design (b) electrode profile (c) sensor mounted on a glass plate

affecting the FRF expression of the electric charge Q_i. Everything else is the same. Figure 26 compares the piezoelectricity profile of the sensor in the two cases tested; note that the overall sensitivity of the anisotropic sensor is much higher because of the much larger value of d_{31}. The quality of the isotropic sensor has been found very similar to that of the copolymer one.

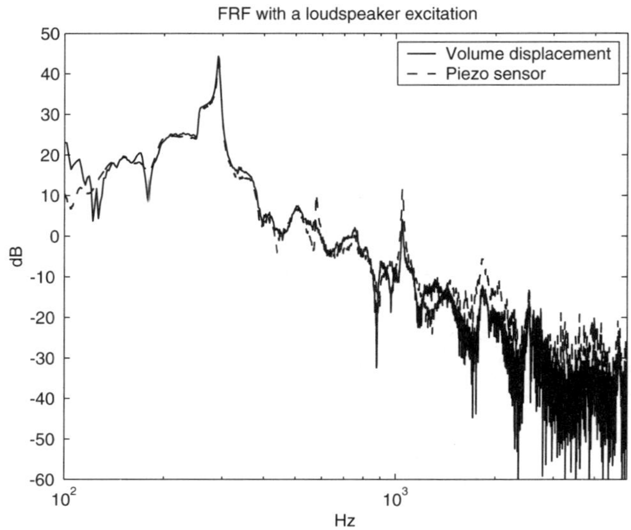

Fig. 25. Transparent (copolymer) PVDF sensor mounted on a glass plate. FRF between the current applied to the loudspeaker and the volume displacement; comparison between the porous piezoelectric sensor (dotted line) and the laser vibrometer (full line)

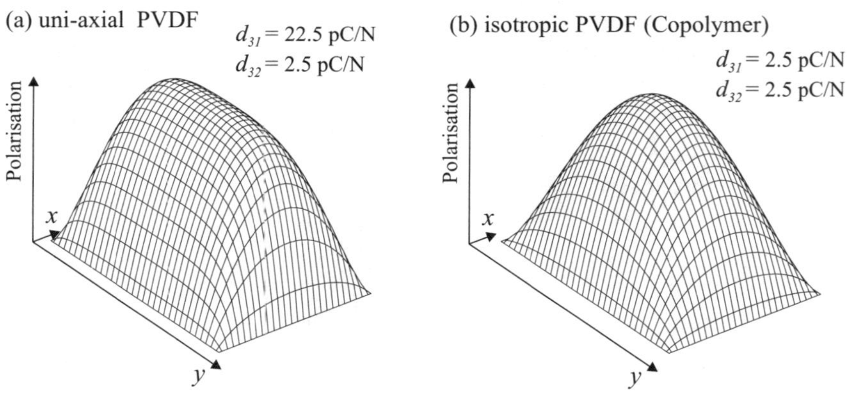

Fig. 26. Comparison of the polarization profile of the volume displacement sensor for various piezoelectric properties

10 Conclusion

This paper has investigated the construction of spatial filters with either discrete array sensors, or with continuous PVDF films. The use of a programmable linear combiner for modal filtering and shaping of FRFs has been addressed. The spatial aliasing of discrete arrays has been discussed. A porous electrode has been discussed, which allows to tailor the effective piezoelectric properties in two dimensions. The various steps of the design procedure have been described and validated with several experiments. Some technological issues such as the use of uni-axial PVDF and a transparent implementation have also been addressed.

Acknowledgement. ULB's contribution in this work was supported by the Ministry of Region Wallonne (DGTRE) under grant n°014427 (SAAB) and the Inter University Attraction Pole IUAP 5 on Advanced Mechatronic Systems. Micromega Dynamics was partly supported by the Ministry of Region Wallonne (VITAC) and by ESA/ESTEC.

References

1. Meirovitch L, Baruh H (1985) The implementation of modal filters for control of structures. AIAA Journal of Guidance, 8(6):707–716
2. Zhang Q, Allemang RJ, Brown DL (1990) Modal filter: Concept and applications. pp.487–496 International Modal Analysis Conference
3. Shelley SJ, Lee KL, Aksel T, Aktan AE (1995) Active control and forced-vibration studies on highway bridge. J. Struct. Eng., pp.1306–1312, September
4. Tanaka N Kikushima Y (1999) Active modal control and its robustness using point sensors and point actuators. JSME Int. Journal, series C 42(1):54–61
5. Sumali H, Meissner K, Cudney HH (2001) A piezoelectric array for sensing vibration modal coordinates. Sensors and actuators A 93:123–131
6. François A, De Man P, Preumont A (2001) Piezoelectric array sensing of volume displacement : A hardware demonstration. J. Sound Vibration, 244(3):395–405
7. Collins SA, Miller DW, Von Flotow AH (1994) Distributed sensors as spatial filters in active structural control. J. Sound Vibration, 173(4):471–501
8. Miu DK (1993) Mechatronics: Electromechanics and Contromechanics. Springer-Verlag, New-York
9. Burke SE, Hubbard JE (1987) Active vibration control of a simply supported beam using a spatially distributed actuator. IEEE Control Systems Magazine, pp.25–30
10. Lee CK (1990) Theory of laminated piezoelectric plates for the disign of distributed sensors/actuators. Part I : Governing equations and reciprocal relationships. J. Acoust. Soc. Am. 87(3):1144–1158
11. Lee CK, Moon FC (1990) Modal sensors/actuators. ASME J. Appl. Mech., 57:434–441
12. Gu Y, Clark RL, Fuller CR, Zander AC (1994) Experiments on active control of plate vibration using piezoelectric actuators and polyvinylidene fluoride (pvdf) modal sensor. J. Vibration and Acoustics, 116:303–308

13. Clark RL, Burke SE (1996) Practical limitations in achieving shaped modal sensors with induced strain materials. J. Vibration and Acoustics, 118:668–675, October
14. Charette F, Berry A, Guigou C (1998) Active control of sound radiation from a plate using a polyvinylidene floride volume displacement sensor. J. Acoust. Soc. Am., 103(3):1493–1503
15. Tanaka N, Snyder SD, Hansen CH (1996) Distributed parameter modal filtering using smart sensors. Journal of Vibration and Acoustics, 118:630–640
16. Rex J, Elliott SJ (1992) The QWSIS - a new sensor for structural radiation control. In: MOVIC, Yokohama, pp.339–343
17. Kim J, Hwang J-S, Kim S-J (2001) Design of modal transducers by optimizing spatial distribution of discrete gain weights. AIAA Journal, 39(10):1969–1976
18. Miller SE, Oshman Y, Abramovich H (1996) Modal control of piezolaminated anisotropic rectangular plates, part 1: Modal transducer theory. AIAA Journal, 34(9):1868–1875
19. Preumont A, François A, De Man P, Piefort V (2003) Spatial filters in structural control. J. Sound and Vibration, 256(1):61–79
20. Preumont A, François A, De Man P, Loix N (2002) A novel electrode concept for spatial filtering with piezoelectric films : Experimental validation. In: Photonics Fabrication Europe SPIE Conference, 28 October - 1 November 2002, Brugge, Belgium
21. Preumont A, François A, De Man P, Loix N (2003) Spatial filtering for active vibration control of plates and shells. In: 5th European Conference on Noise Control, 19-21 May 2003, Naples, Italy
22. Preumont A, François A, De Man P, Loix N, Henrioulle K. Spatial filters in structural control part II : Distributed sensors with piezoelectric films. Submitted to the J. Sound Vibration
23. Gawronski WK (1998) Dynamics and Control of Structures, a Modal Approach. Springer
24. Strang G. (1988) Linear Algebra and its Applications, 3rd Ed. Harcourt Brace Jovanovich
25. Martin GD (1978) On the Control of Flexible Mechanical Systems. PhD thesis, Stanford University
26. Piefort V (2001) Finite Element Modeling of Piezoelectric Active Structures. PhD thesis, Université Libre de Bruxelles, The software is currently part of SAMCEF^{TM} of SAMTECH S.A. Liège - Belgium
27. Johnson ME, Elliott SJ (1995) Active control of sound radiation using volume velocity cancellation. J. Acoust. Soc. Am., 98(4):2174–2186
28. Preumont A (2002) Vibration Control of Active Structures, An Introduction. Kluwer Academic Publishers, Dordrecht, second edition

Structural Health Monitoring Using Guided Ultrasonic Waves

Wiesław J. Staszewski

Department of Mechanical Engineering
Sheffield University
Mappin Street, Sheffield S1 3JD, United Kingdom
w.j.staszewski@sheffield.ac.uk

1 Introduction

Maintenance of air, land and sea structures is an important engineering activity in a wide range of industries including transportation and Civil Engineering. Effective maintenance minimises not only the cost of ownership of structures but also improves safety and the perception of safety. Inspection for material/structural damage, such as fatigue cracks and corrosion in metallics or delamination in composites, is an essential part of maintenance.

The last few decades have created an increasing need for reliable, early detection of damage in structures. Various experimental techniques, approaches and methodologies have been developed. Recent years have shown applications of guided ultrasonic waves and smart sensors which can become an integral part of a monitored structure. This includes damage detection techniques based on Lamb waves. These techniques are successful in laboratory conditions and have great potential for field applications.

For simple laboratory structures, damage detection based on Lamb waves is not difficult. However, complex, real engineering structures require appropriate monitoring strategies and signal processing for successful damage detection. Using Lamb waves for damage detection requires some experience and caution. It appears that significant theoretical and experimental experience have been gathered in this area over the last twenty years. This chapter collates in one place recent developments related to structural damage detection based on guided ultrasonic waves. The focus is on sensor technologies, various aspects related to monitoring strategy, transducer and wave propagation modelling techniques and signal processing procedures for damage detection. It is beyond the scope of this book to discuss all important elements; the reader is directed to appropriate references for further details.

1.1 Elastic Waves

Elastic waves are travelling disturbances that transport energy in a elastic medium without transferring matter. Materials are considered as elastic if the energy expended in deforming their elements is stored in these elements only as potential energy which is often called strain energy.

It is well known that various types of waves can propagate in solids, as described in [1]. Waves travelling in a unbounded bulk of elastic material are called bulk waves. These waves have a finite number of longitudinal and shear components. Longitudinal (often called compressional, dilatational pressure or P) waves exhibit particle motion in the direction of wave propagation. The direction of particle motion in shear (often called transverse or S) waves is perpendicular to the direction of wave propagation. Shear waves can be polarised horizontally and vertically leading to SH and SV wave components, respectively. Wave propagation in bounded media leads to guided waves such as plate and surface waves described in Section 2. Guided waves exhibit an infinite number of modes resulting in a complex wave propagation mechanism.

Bulk and guided waves are governed by the same mathematical equations. The fundamental elastodynamic equation, which is based on the linear stress-strain relation and Newton's second law, can be used to describe elastic wave propagation as

$$\rho \ddot{u}_i = \left(s_{ijkl} u_{k,l}\right)_{,i} \tag{1}$$

where ρ is the density, u is the displacement, s is the stiffness tensor and the standard tensor index convention is used. The bulk wave solution of this equation does not need to satisfy boundary conditions. In contrast, the guided wave solution must satisfy physical boundary conditions.

1.2 Damage Detection Using Elastic Waves

Elastic waves and their propagation have been used for many years to analyse structural damage. A number of different techniques based on elastic waves have been developed for Non-Destructive Testing and Evaluation (NDT/E). Acoustic Emission (AE) and Ultrasonic Testing (UT) are the most widely used techniques in industrial applications. Damage detection based on guided ultrasonic waves and acousto-ultrasonic techniques have reached much less maturity but also have shown great potential for Structural Health Monitoring (SHM) applications.

AE [2, 3, 4] utilises transient stress waves, called acoustic events, emitted internally by different types of material defects under external loading applied to monitored specimens. Material defects, such as microcracks and/or inclusions, release elastic energy due to rapid local stress redistribution as a result of loading. The energy results from growing cracks, rubbed surfaces of cracks, dislocations, phase transformation or melting. The frequency of acoustic events is usually between 100-1000 kHz.

UT [1, 5] is the most versatile damage detection technique based on elastic wave propagation. The method utilises various phenomena and/or properties of ultrasonic waves propagating in the material to detect defects. It relies predominantly on the transmission and reflection of high-frequency ultrasonic bulk waves. The frequency of these waves is usually in the range from 40 to 2000 kHz and can go far beyond this bandwidth in the case of acoustic microscopy. Ultrasonic testing utilises wave attenuation, scattering, reflections, mode conversions and energy partitioning for damage detection.

Lamb wave inspection is the most widely used damage detection technique based on guided ultrasonic waves. The first NDT/E application of Lamb waves goes back to the 1950s [6]. The technique is based on Lamb waves propagating in plate-like structures. These waves are introduced by a probe at one point and sensed by another probe at a different position, as illustrated in Figure 1 [1]. Structural damage is identified by a change of the response signal. Often wave attenuation and/or mode conversions are sufficient to detect defects.

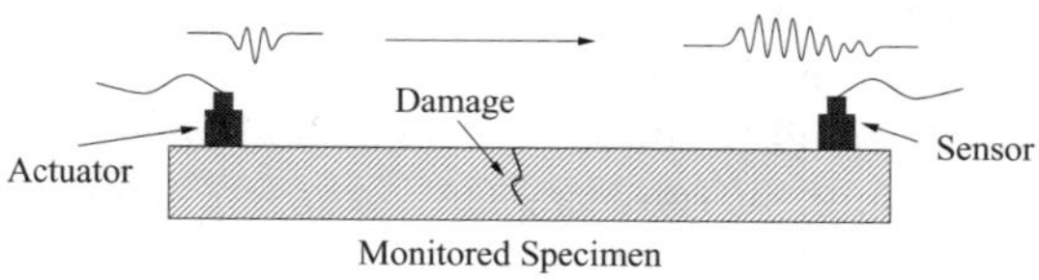

Fig. 1. Structural damage detection based on guided ultrasonic waves.

The acousto-ultrasonic technique was introduced in the late 1970s and reported in [9, 10]. This technique uses high-frequency (usually above 0.5 MHz) impulse excitation which results in a large number of mixed propagating wave modes. Alternatively, other types of broadband excitations such as, swept sine waves and Gaussian white noise, can also be used [11]. The rich energy content of propagating waves includes lots of information regarding possible material and/or structural damage.

The acousto-ultrasonic technique combines elements of AE, UT and guided ultrasonic waves. All these four techniques utilise stress waves propagating in structures for damage detection. However, AE is the only technique which does not require any external signal excitation; stress waves are structure-borne and produced by material defects when external loading is applied. It is a passive technique. The other three techniques require high-frequency external excitation. Wave propagation paths in UT and damage detection techniques based on guided ultrasonic waves are well-defined and traceable. In contrast, wave propagation in acousto-ultrasonics is difficult to analyse

[1] In practice the procedure can utilise only one transducer. Reflection signals are then used for damage detection, as explained in Section 4.5. This approach requires the use of a full bridge circuit for Lamb wave generation and sensing [7, 8].

since transducer responses include a number of mixed modes not only from directly propagating waves but also from reflected and scattered waves.

2 Guided Ultrasonic Waves

2.1 Background

Wave packets propagating in bounded media, which are superpositions of various modes, are often called guided waves. Guided waves are governed by the same wave equations as bulk waves. However, in contrast to bulk waves, they have an infinite number of modes associated with propagation. There are various types of guided waves. Wave packets, resulting from appropriate stress and strain boundary conditions, which travel on the surface of a solid body, are known as surface waves. Surface waves usually exhibit large amplitudes and travel slower than other types of guided waves. Rayleigh waves, which are boundary waves in an elastic half space, are the best known surface waves. They are nondispersive for uniform material properties. However, their mechanism of propagation is very complex; waves are polarised and surface particles are moved around in an ellipse. The components of Rayleigh waves can couple with a medium surrounding the surface of the body. This coupling affects the amplitude and velocity of the wave. The amplitude of the wave decreases rapidly with depth and the rate of decrease depends on the wavelength. Therefore inspection methods based on Rayleigh waves are used mostly to detect surface defects. Other examples of surface waves include Stonely and Love waves. These waves are not commonly used for damage detection and are better known to seismologists than to damage detection experts. Stonely waves occur at an interface between two media. They are closely related to Rayleigh waves. The existence of Stonely waves depends on the density and shear modulus ratios of the neighbouring media. Love waves are horizontally polarized shear waves which also exist on the surface. However, in contrast to the other two types of surface waves they are highly dispersive. Love waves are the fastest propagating surface waves. Lamb waves are dispersive plate waves that occur for traction-free forces on both surfaces of the plate. The waveguide of these waves is defined by two boundaries, i.e. the top and bottom of the plate.

For various modes the velocity of these waves depends on the product of frequency of excitation and thickness of the plate. They can propagate long distances and are used for damage detection in plate-like structures. Lamb waves are the most widely used guided waves for damage detection.

2.2 Lamb Waves

Lamb waves refer to elastic perturbations propagating in a solid plate with free boundaries. This problem was firstly studied in [12] and is also very well described in [1, 13]. A brief introduction to the theory of Lamb waves is given in this section.

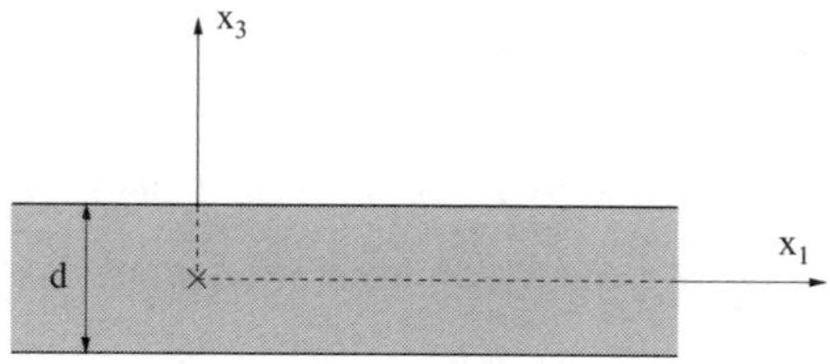

Fig. 2. Wave propagation coordinates for Lamb wave equations.

Figure 2 shows the geometry of the analysed plate of thickness d. Here x_1,x_2 and x_3 indicate the directions of displacement. When the plate is excited at some point, Lamb waves arise from a coupling between propagating shear and longitudinal waves reflected at the top and bottom of the plate. The wave propagation can be described using Equation (1). The solution of this equation can be obtained using different methods. This includes the displacement potentials approach [14] and the partial wave technique [15]. Following [1] it can be shown that the method of potentials transforms displacements into field variables ϕ and ψ using the Helmholtz decomposition. As a result, the elastodynamic wave equation becomes uncoupled and for plane strain takes the form of two independent equations

$$\frac{\partial^2 \psi}{\partial x_1^2} + \frac{\partial^2 \psi}{\partial x_3^2} = \frac{1}{c_L^2} \frac{\partial^2 \psi}{\partial t^2} \tag{2}$$

$$\frac{\partial^2 \phi}{\partial x_1^2} + \frac{\partial^2 \phi}{\partial x_3^2} = \frac{1}{c_T^2} \frac{\partial^2 \phi}{\partial t^2} \tag{3}$$

where c_L indicates the velocity of longitudinal waves whereas c_T is the velocity of shear waves. It is clear that Equations (2) and (3) describe propagation of longitudinal and shear waves, respectively. Physical displacements of the plate can be expressed in terms of the field variables as

$$u_1 = \frac{\partial \psi}{\partial x_1} + \frac{\partial \phi}{\partial x_3}$$

$$u_2 = 0 \tag{4}$$

$$u_3 = \frac{\partial \psi}{\partial x_3} - \frac{\partial \phi}{\partial x_1}$$

Final solutions for displacement are often split into symmetric and antisymmetric modes due to the fact that the field variables involve sine and cosine functions which are odd and even functions, respectively. The displacement for the symmetric modes can be given as

$$u_1 = ikC_2cos(px_3) + qD_1cos(qx_3) \tag{5}$$

$$u_3 = -pC_2sin(px_3) - ikD_1sin(qx_3) \tag{6}$$

whereas the solutions for the antisymmetric modes are

$$u_1 = ikC_1sin(px_3) - qD_2sin(qx_3) \tag{7}$$

$$u_3 = pC_1cos(px_3) - ikD_2cos(qx_3) \tag{8}$$

The variables p and q in Equations (5-8) are given as

$$p^2 = (\omega/c_L)^2 - k^2; \qquad q^2 = (\omega/c_T)^2 - k^2 \tag{9}$$

where k is wave number and $\omega = 2\pi f$ is circular frequency. The constants C_1,C_2,D_1 and D_2 in Equations (5-8) can be obtained from the traction-free boundary conditions for the analysed plane strain. This leads to Rayleigh-Lamb frequency relations

$$\frac{tan(qh)}{tan(ph)} = -\frac{4k^2pq}{(q^2-k^2)^2} \tag{10}$$

for symmetric modes and

$$\frac{tan(qh)}{tan(ph)} = -\frac{(q^2-k^2)^2}{4k^2pq} \tag{11}$$

for antisymmetric modes, where $h = d/2$. These equations can be solved numerically as explained in [1]. The results give either values of phase velocity c_p or group velocity c_g ($c_p = \omega/k$ and $c_g = d\omega/dk$) as a function of the fd (frequency $\times$ thickness) product. Figure 3 gives an example of the phase velocity dispersion curve for an aluminium plate. Here, each curve represents a specific mode, which is conventionally called S_0, A_0, S_1, A_1, S_2, A_2, etc., where S_n and A_n ($n = 1, 2, 3...$) denote symmetric and antisymmetric modes, respectively. For $fd < 1.9$ only the fundamental S_0 and A_0 can propagate in the aluminium plate. Dispersion characteristics can be obtained for any type of material. The software suite *Disperse* [16, 17] has been developed for dispersion characteristics in multilayered media. Also, the FE method has been employed to obtain mode shapes and guided wave characteristics for structures with complex geometries [18]. It is important to mention that Lamb waves arise from coupling between longitudinal and shear waves; the S_0 mode is dominated by the longitudinal component whereas the A_0 mode is dominated by the shear component.

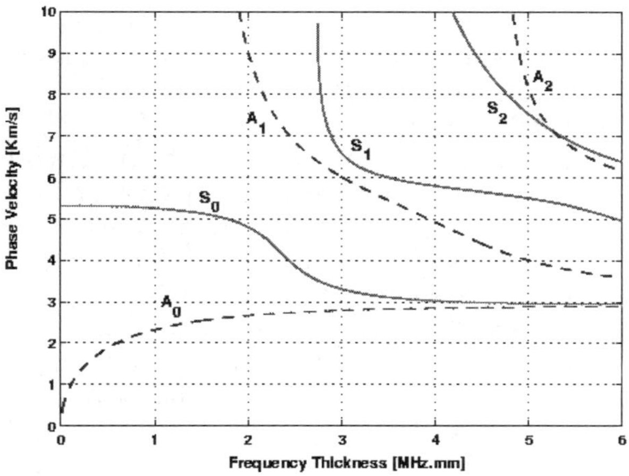

Fig. 3. Phase velocity dispersion characteristic for aluminium.

3 Generation and Sensing - Transducers

Ultrasonic and guided ultrasonic waves can be generated and sensed utilising various physical phenomena, as illustrated in Figure 4. All these transducers can be classified into two major groups. The first group uses perturbations on the surface of the structure (e.g. piezoelectric-based transducers) whereas the second group utilises perturbations inside the structure (e.g. electromagnetic transducers). Another possible classification of transducers generating/sensing guided ultrasonic waves leads to contact and non-contact techniques. A brief summary to different types of transducers is given in this section.

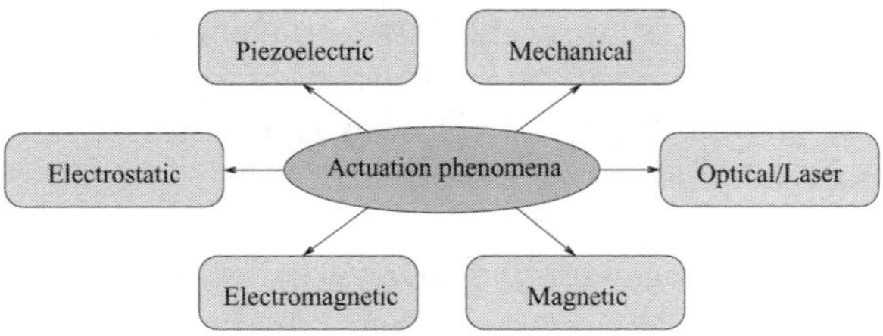

Fig. 4. Actuation phenomena used in transducers generating and sensing guided ultrasonic waves.

3.1 Contact Techniques

Sensor and actuators utilising piezoelectric materials are the most widely used transducers for damage detection based on Lamb waves. Piezoelectricity is an electric polarisation effect due to mechanical forces, i.e. an electric charge is collected on the surface of the piezoelectric material when it is squeezed. Often the converse effect is possible, i.e. the material generates a mechanical strain in response to the applied electric field.

Piezoelectricity was discovered by Jacque and Pierre Curie in the 1880s. The historical perspective of theories, conceptual models and experiments related to this physical effect is given in [19]. Piezoelectricity is an anisotropic property of crystalline materials and results from non-uniform charge distributions within the crystal's cells. There are number of different materials which exhibit these effects. Natural materials include: quartz (SiO_2), Rochelle salt and tourmaline. The quartz crystal exhibits stiffness comparable to steel and shows high voltage sensitivity. Manufactured piezoelectric ceramics were introduced in the early 1950s. The first material developed with piezoelectric properties was barium titanate ($BaTiO_3$). The two most widely used manufactured materials are: lead zirconate titanate (formally known as PZT) and polyvinylidene fluoride (formally know as PVDF). The former is a ceramic and the latter is a polymer film. Piezoelectric ceramics are quite brittle and need to be handled with care. Piezoelectric polymer films are in contrast very flexible and easy to handle for sensor applications. The PVDF exhibits the strongest known piezoelectric behaviour of all polymers. The PVDF offers better direct piezoelectricity and worse inverse piezoelectricity than the PZT and therefore it is more often used for sensing applications. An excellent tutorial on piezoceramics is given in [20, 21]. Table 1 gives a summary of various properties for some commercially available piezoceramics used for damage detection based on Lamb waves.

For many years conventional ultrasonic transducers utilising piezoelectric elements have been used for Lamb wave experimental analysis. This requires either appropriate water coupling [22, 23] or a perspex angled wedge (e.g. [24, 25, 26, 27]). Transducers utilising the wedge are better know as wedge (or angle-beam) piezoelectric transducers. These transducers use piezoelectric elements which are positioned with an angle to the entry surface of a monitored structure, as illustrated in Figure 5a. The angle and the size of the transducer determine the wavenumber bandwidth.

Recent years have shown extensive progress in developing low-profile piezoelectric sensing devices. The adaptation and integration of piezoelectric sensors onto, or into, structures has become more feasible. Low-profile piezoelectric sensors with wrap-around electrodes are available in the form of thin ceramic plates and discs which can be bonded (or embedded) on monitored structures. These sensors are commonly known as piezoceramics. Other important developments in this area include piezoceramic rubbers and paints [28], *Smart Layers*^{TM} [29], smart hybrid panels [30],piezoceramic composites [31, 32] and piezoceramic fibres [33]. Rubbers and paints are loaded with a fine powder of piezoceramics. *Smart Layers*^{TM} comprise small piezoceramic discs on a thin dielectric Kapton layer which can be easily surface mounted

Table 1. Properties of typical piezoceramics used for transducer applications; values taken from manufacturer's data sheets.

Material		PZT-5A	PZT-5H	Sonox-P5	P-155	C5500	PZT
Manufacturer		Piezo Systems		CeramTec	PI Ceramics	Channel Industries	Acellent
Relative dielectric constant	$\frac{\epsilon_{33}^{T}}{\epsilon 0}$	-	-	1850	1500	-	1750
Piezoelectric constants 0^{-12} m/V	d_{33}	390	650	450	360	400	400
	d_{31}	-190	-320	-180	-165	-185	-175
Piezoelectric constants 10^{-3} Vm/N	g_{33}	24	19	27.5	27	25.8	26
	g_{31}	-11.6	-9.5	-	-12.4	-11.9	-12.4
Coupling Coefficients	k_p	-	-	0.62	0.62	0.62	0.63
	k_{33}	0.72	0.75	0.73	0.69	0.73	0.72
	k_{31}	0.32	0.44	0.34	0.35	0.37	0.36
Curie temp., ^{o}C	T_C	350	250	340	345	350	360
Elastic modulus 10^{10} N/m^2	Y_{11}^E	6.6	6.2	5.8	6.4	6.4	6.3
	Y_{33}^E	5.2	5.0	5.3	4.8	5.2	5.4
Mechanical quality factor	Q_m	80	30	90	80	75	80
Density, g/cm^3	ρ	7.8	7.8	7.65	7.8	7.6	7.7

on a structure or integrated into a structure. These layers can be fabricated in various sizes and shapes, as shown in [34]. Smart hybrid panels utilise *Smart Layers*^{TM} and optical fibre sensors.

Electro-Magnetic Acoustic Transducers (EMAT) (e.g. [35, 36]) are also widely used for Lamb wave experiments. EMATs are electromagnetic devices that consist of a magnet (coil) and a stack of wires. EMAT transducers generate ultrasonic waves in electrically conductive materials by Lorentz force, electro-magnetostrictive effect or by a combination of these two phenomena. These transducers do not require any couplant. The wavenumber bandwidth of the EMAT transducer is a function of the number of limbs on the excitation coil.

The excitation of fundamental Lamb wave modes over a controlled frequency bandwidth can significantly ease damage detection. This can be achieved using a comb transducer (e.g. [1, 37]) or its modern version called an interdigital transducer (e.g. [38, 39]). Typical comb transducers are constructed as corrugated metal plates excited by a piezoelectric patch. The desired wavelength is controlled by the ridge spacing. Interdigital transducers have a pattern of electrodes deposited on a piezoelectric substrate, as shown in Figure 5b. The wavelength is controlled by the space between the finger electrodes.

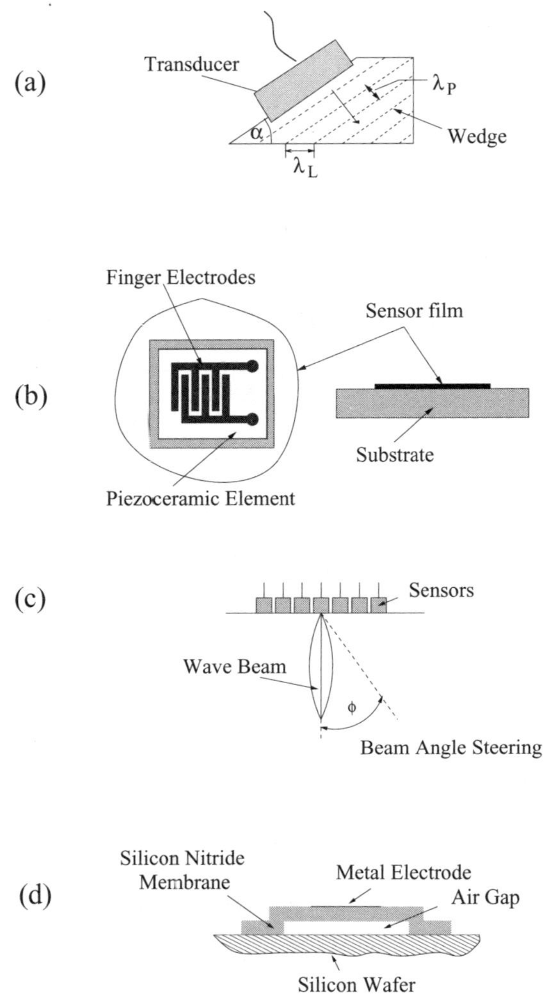

Fig. 5. Example of transducers used for ultrasonic guided wave generation: (*a*) angle-beam (*b*) interdigital (*c*) phased array (*d*) MEMS.

Phased array transducers (e.g. [40, 41, 42]) generate Lamb waves propagating in various directions controlled electronically (Figure 5c. The angle and the focal length of propagation of these transducers can be adjusted by appropriate delays of the signals emitted by an array of small crystal elements. These elements can work in a combined transmitter/receiver mode [43] allowing for a significant reduction of transducers required for damage detection.

Recent advancements in the area of Lamb wave generation/sensing includes applications of Micro-Electro Mechanical Systems (MEMS) [44, 45] and Bragg grating optical fibre sensors [46, 47]. MEMS transducers used in

[44, 45] are capacitive micromachined ultrasonic transducers (CMUTs). These transducers are in fact thin silicon nitride membranes supported on 500 μm silicon substrates, as illustrated in Figure 5d. Lamb waves are generated by vibrating membranes. The frequency of CMUTs transducers is in the vicinity of 1 MHz. The basic idea of Lamb wave interrogation using optical fibre Bragg grating sensors is to use a laser with a narrow bandwidth. If the wavelength of the laser matches a certain part of the grating spectrum, any shift of the spectrum will as a consequence modulate the reflected optical power. The major advantage of the Bragg grating sensors is the ability to monitor both static and dynamic strain fields. Load monitoring and damage detection are possible at several gratings within a single fibre.

3.2 Non-Contact Techniques

Non-contact generation and reception of Lamb waves can be achieved using air-coupled transducers (e.g. [48, 49]) various types of laser/optical techniques such as homodyne, heterodyne, Fabry-Perot and photorefractive interferometry [50]. Air-coupled transducers have very low sensitivity due to the large acoustic impedance mismatch between air and solid materials and due to the high attenuation of ultrasound in air (dry couplant) at frequencies above 500 kHz. Laser generation of Lamb waves utilise high power laser sources [51, 52] or low-power modulated laser diode [53] Damage detection application examples of laser-generated Lamb waves include [51, 54, 55, 56, 57, 58]. The high peak optical powers associated with pulsed laser systems (e.g. Nd:YAG) can be utilised to provide a short rise-time and spatially localised expansion source in samples, which in turn serves as a source of broadband elastic stress waves [59]. High power laser energy focussed over a small area leads to two phenomena: thermoeleastic expansion (incident power below $10^7 W/cm^2$) and ablation (incident power above $10^7 W/cm^2$; the source works in plasma regime), as explained in [51].The former is nondestructive; ultrasound is generated due to constrained local rapid expansion. The latter is destructive; ultrasound is generated due to momentum expansion.

Optical detection of ultrasound can be performed in a variety of ways. Commonly used detectors measure changes in optical intensity. In optical beam deflection (e.g. [54, 55]), the intensity change is measured directly. The detected elastic wave modulates the surface and changes the amount of light that reaches a photodetector. In the interferometric techniques (e.g. [56, 60, 61]) the elastic wave modulates the optical phase of the light. The interferometer is then used to demodulate the signal into an intensity fluctuation monitored by a photodetector. The most commonly used interferometric detectors are Fabry-Perot, Michelson and Mach-Zehnder. A two-channel interferometer (Mach-Zehnder and Michelson) has been used for Lamb wave detection in [62]. Laser interferometers can be designed to measure either out-of-plane or in-plane motion. Some interferometers are sensitive to surface displacement, whilst others are made sensitive to surface velocity. A velocity sensitive system (e.g. heterodyne Michelson) is less susceptible to low frequency environmental vibrations which is a major practical advantage. However, the Fabry-Perot configuration is often championed because of its superior operation in real engineering materials in terms of its light gathering power. All these practical aspects are discussed in more details in [63, 64].

The major disadvantage associated with Lamb wave sensing using laser interferometers is the high cost, the difficulty of the appropriate alignment and safety issues. Optical beam deflection systems are considerably less expensive, easier to operate but less sensitive.

Laser vibrometers are interferometers which measure the out-of-plane surface motion utilising the Doppler shift phenomenon in order to obtain the velocity of the surface vibration. They are designed and widely used for vibration measurements in many engineering applications. This includes modal analysis [65, 66, 67] and vibration-based damage detection [68, 69]. Commercial laser vibrometers have also been used to analyse the stress-acoustic field [70], ultrasonic shear waves [71] and Lamb waves [72, 73, 74, 75].

Lamb wave generation and sensing can be also accomplished using another dry-contact, point-like method utilising an elastic Hertzian contact between solid bodies, as demonstrated in [76]. The Hertzian contact between specially shaped buffer rods (metallic or quartz) and specimens allows for direct energy coupling, i.e. displacements induced by extensional waves in the rods are coupled to Lamb wave modes in the specimen.

4 Monitoring Strategy for Damage Detection

The subtleties of Lamb wave with the complicated physics require some experience related to monitoring strategy and complex data interpretation. The basic factors which determine damage detection based on Lamb waves are related to [77, 78, 79, 80, 81]: properties of the structure under inspection, selection of wave propagation modes, transducer selection and location, transducer validation against possible failures, transducer coupling methods and choice of excitation signals, as briefly discussed in this section. Other elements, described in the next section, include various aspects related to signal processing techniques for damage feature extraction.

4.1 Transducer Schemes

Selection of appropriate transducers is important for successful damage detection. Contact transducers require coupling (gel, water) for the ultrasound energy transfer. Special types of transducers are used when specimens are immersed in water. Coupling significantly attenuates ultrasonic energy induced from the transducers to the specimen, as demonstrated in [82]. Non-contact transducers do not require any coupling. They are very attractive for scanning and are capable of monitoring large surfaces. However, non-contact transducers show significant attenuation at high frequencies, as already mentioned in Section 3.1. Permanently attached transducers can be either bonded on the surface of monitored structures or embedded into the structures. Low-profile piezoceramic transducers (e.g. *Smart Layers*^{TM}) and optical fibre sensors are often embedded in composite structures. Although embedded transducers offer many advantages, it is difficult (if not impossible) to replace them in case of functional failures.

Modal selectivity and transducer directivity are important elements associated with Lamb wave generation. Although theoretically a minimum two Lamb wave modes always propagate in the analysed structure, it is possible to chose excitation frequencies for which the amplitude of one fundamental mode is reduced almost to zero. This is often called single-mode propagation. Single-mode propagation simplifies damage detection analysis of propagating Lamb wave modes. Good spatial and temporal generation characteristics are required from transducers chosen for single-mode propagation. Wedge piezoelectric transducers and multi-element piezoceramic transducers (e.g. comb or interdigital transducers) offer good spatial generation characteristics [78]. Optimal temporal excitation can be achieved with appropriate excitation signals are used, as discussed in Section 4.3. It is well known that some piezoceramics (e.g. SONOX-P5) elements are better to generate symmetric modes wheras other piezoceramics (e.g. P155) are more appropriate for antisymmetric mode excitation. Multi-mode generation, often used in acousto-ultrasonics, can be accomplished using broadband excitation offered for example by pulsed laser excitation.

The shape of piezoceramic transducers determines the directivity of wave propagation, as shown in [81]. Rectangular piezoceramics offer parallel generation/propagation of waves to transducer edges whereas circular piezoceramics generate and propagate waves circumferentially. The directivity of transducers can be roughly estimated using the relationship between the length (diameter) of the transducer b and the actuating frequency f [13]

$$b = \lambda \left(n + \frac{1}{2} \right) = \frac{c_p}{f} \left(n + \frac{1}{2} \right) \tag{12}$$

where λ is the wavelength and $n = 0, 1, 2, 3, ...$ This equation estimates the desired dimension of the transducer in the direction of wave propagation for most efficient response signals.

Successful damage detection also requires understanding of the ultrasonic field of transducers. This involves problems related to pressure profiles of transducers, near-field calculations and beam angle of divergence, as discussed in [1]. The analysis of transducer beam pressure profiles in [1] shows that collimated beams can be obtained for $b/\lambda > 3$ and circumferential wave propagation is possible for $b/\lambda < 0.5$. The study in [83] indicate that the width of a transducer should be five to ten times the wavelength of Lamb waves in order to generate well collimated beams.

Recent advancements in transducer technologies have shown application of wireless and multifunctional transducers. Wireless applications are possible with piezoceramic, MEMS, interdigital and phased array sensors, as shown in [84]. MEMS and optical fibres offer multifunctional transducers capable to monitor not only ultrasound but also corrosion, temperature, strain and/or manufacturing processes. Application examples include [29, 46].

4.2 Transducer Location and Validation

SHM systems based on Lamb wave inspection requires appropriate location of transducers for reliable damage detection. Deciding the optimal positions of transducers is often based on engineering knowledge, expertise and experience. In practice, designers and end-users of monitored structures know where the critical areas are. When direct responses are used for damage detection, actuator and sensor should be located in a line-of-sight position regarding the damage. If the damage is located far away from the actuator-sensor line, Lamb wave sensitivity to damage is reduced, as observed in [128, 85]. Also, damage orientation against the actuator-sensor line is important; the studies in [85] show that damage detection is more reliable for fatigue cracks positioned at the angle of 45^0 rather than 90^0 to the direction of wave propagation in metallic plates. Actuators positioned very close to boundaries (e.g. plate edges) exhibit more directivity due to canceled undesired reflections.

The distance between pairs of transducers should be chosen to avoid large amplitude attenuation which leads to poor signal-to-noise ratio. Lamb waves can propagate long distances in metallic structures. This is not the case in composite structures where the level of attenuation is much more significant. Nevertheless distances up to 1m are still visible for these structures, as shown in [129]. Also, long propagation distances result in more severe dispersion.

A number of signal processing tools have been developed to support the sensor location problem. These tools are often based on optimisation procedures and information theory. It appears that the sensor location problem has been well addressed in control engineering and system/modal identification. However, relatively little experimental and applied research has been done in the area of damage detection. An overview of optimal sensor location methods for damage location is given in [86] where various methods such as classical optimisation techniques (e.g. Newton methods, linear and nonlinear programming),soft computing techniques (e.g Genetic Algorithms (GAs) [87], Simulated Annealing (SA) [88], Tabu Search (TS) [89, 90, 91]) and mutual information [92] have been discussed. Application examples related to damage detection include: the heuristic optimisation approach based on a simple stepwise technique and neural networks [93] a combined neural networks/genetic approach [94, 95] and a mutual information based algorithm [96, 97].

The reliability of transducer readings, or in other words the detection of transducer which have failed and do not measure accurately, is one of the most important problems in many areas of engineering. This problem is particularly important in automated damage detection systems where it is crucial to distinguish between structural damage and transducer faults in order to minimise false-positive (damage detected when none is present) and false-negative (damage not detected when present) situations.

Failures of low-profile piezoceramic transducers used for Lamb wave inspection are mostly related to bonding and connections. Two types of bonding, i.e. cold (using epoxy-based glues) and hot (using autoclave), are used in practice. Transducer connections can be improved using either piezoceramic discs with wrap-around electrodes or *Smart Layer*^{TM} sensors.

Various algorithms for signal and sensor validation have been developed over the last twenty years. Sensor validation techniques can be classified into three basic categories [98]: qualitative methods, physical model-based techniques (parity space [99, 100], observer-based [101] and parameter identification [102] approaches) and data driven models (sequential probability ratio test [103], Auto-Associative Neural Networks - AANN [104, 105], kernel-based techniques [106]). Sensor validation procedures for damage detection and smart structures have been discussed in [107, 108].

4.3 Mode Selection

Lamb wave mode selection is also important for damage detection. The choice of fundamental modes reduces the complexity of the propagating wave. This is particularly relevant to the single-mode approach, as discussed in Section 4.1. Both fundamental modes are very good in terms of the spatial resolution, i.e. good compromise between the bandwidth of the excitation signal and its duration, as discussed in [77]. The symmetric S_0 mode is faster and has much lower attenuation than the antisymmetric A_0 mode. It is also less dispersive. Generally, it is advantageous to excite modes in non-dispersive regions since dispersion leads to the increase of signal duration (worsens the signal resolution) and the reduction of signal amplitude (limits the propagation distance) [109], as further discussed in Section 4.4.

Although, the S_0 mode is considered as less sensitive to detect damage, a number of studies have been reported where this mode works better than the A_0 mode. Examples include weld defects in metallic plates [109] and delaminations [110], thermal damage [111] in composite materials.

The sensitivity of Lamb wave modes to different types of defects also needs to be considered in practical damage detection applications, as discussed in [27]. The energy distribution for various modes is different through the thickness of the monitored plate. This is due to the fact that the in-plane and out-of-plane displacements vary across the thickness, as illustrated in [27]. In general, lower modes have more uniform energy distribution than higher modes. However, the S_0 has a more uniform distribution than the A_0 mode which exhibits higher levels of energy at the surface. This suggest that the S_0 mode is a safe choice for any type of defects and the A_0 mode is more suitable for surface cracks or corrosion. However, some studies (e.g. [112]) indicate that the S_0 mode is not good to detect delaminations in composite materials. Numerical defect depth studies have been performed in [76, 113, 114] to analyse the potential of using various Lamb wave modes for damage detection. These studies show that the S_0 mode is more sensitive to a flaw or inclusion located in the centre of the material whereas the A_1 mode is more sensitive to a surface crack.

4.4 Excitation Signals

The dispersive nature of Lamb waves and the large number of possible modes at a given frequency makes the inspection very difficult. Thus it is important to limit the bandwidth of excitation to a range over which there is little dispersion and to focus the analysis on fundamental modes. Firstly, single-mode excitation offered by some types of transducers, as already discussed in Section 4.1, should be used. Secondly, points of stationary group velocities from the Lamb wave dispersion curves should be used for the selection of excitation frequencies. Thirdly, the bandwidth of frequency excitation should be as narrow as possible. However, a compromise needs to be achieved between the bandwidth and length of the excitation signal since it is easier to analyse Lamb wave responses which are as short as possible. This is due to the fact that short signals allow for good temporal separation, i.e. incident and reflected waves can be more easily separated. In practice tone burst windowed signals with a large number of cycles are used as an excitation in order to limit the bandwidth. Figure 6 gives an example of the excitation signal used for Lamb wave inspection. The number of cycles needs to be optimised in order to obtain a response wave as short as possible. All these aspects are discussed in [77].

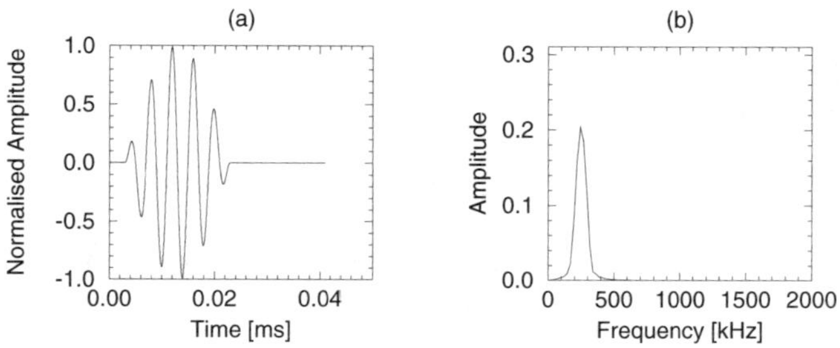

Fig. 6. Example of excitation signal used for damage detection based on Lamb waves: (*a*) time domain (*b*) frequency domain.

The frequency of excitation is directly related: to mode selection, as discussed in Sections 2.2 and 4.3, and to dispersion, as explained above. Figure

3 shows that thicker components require lower frequencies for fundamental mode propagation. Broadband excitation (pulse, sweep sine, Gaussian white noise) is used for acousto-ultrasonics, as shown in [11]. If good directivity is required from a given transducer, the frequency needs to be chosen following the guidelines given in Section 4.1.

The maximum performance of the transducer is achieved when the excitation frequency is chosen as one of the natural resonance frequencies of the transducer. This enables a very efficient conversion from the electrical to mechanical energy. Resonance frequencies are associated with different vibration modes of the transducers. These modes can be used for the generation of different types of waves. Vibration-based (e.g. Frequency Response Functions - FRF or transfer functions) and impedance characteristics can be used in order to describe dynamic properties of the piezoceramic transducers, as described in more detail in Section 5.1. These characteristics can be obtained using theoretical and experimental studies.

The excitation frequency is also directly related to the wavelength which in turn affects the sensitivity to damage. The wavelength of the classical ultrasound has a significant effect on the probability of detecting damage. A rule of thumb in ultrasonic applications is that discontinuities that are larger than one-half the size of wavelength can usually be detected. However, it is well known that the sensitivity of Lamb waves is not a simple function of wavelength. Damage detection using guided waves involves more factors than the relationship of wavelength and damage size. The energy of the signal reflected from damage is also dependent on acoustic impedance mismatch between the damage and the surrounding material. It is well-known that cracks slow down waves (velocity is proportional to $\sqrt{E/\rho}$, where E is the modulus of elasticity and ρ is the density). Also, it is important to remember that the wavelength value for short-time pulse waves (pulse wave with a small number of oscillations) varies due to the broadband nature of the pulse signal. The ability of guided waves to detect defects much smaller than the wavelength is widely reported in the literature. Application examples include [115, 116, 117, 118]. The study in [119] shows that the wavelength of S_0 (12 mm) and A_0 (5.7 mm) Lamb wave modes exhibit almost identical amplitudes of their transmission ratios across a 0.5 mm deep notch in a 3 mm thick steel plate. The use of low-frequency Lamb waves for large length scale specimens is demonstrated in [120].

Various energy/power levels of excitation are used for damage detection based on Lamb waves. Lamb wave applications show studies with voltage amplitudes in the range of: 1-20V (PZT transducers - [11, 79, 121]), 100-150V (PZT transducers [34, 122], PDVF transducers [123], angle-beam transducer [112]) and even 500V (air-coupled transducers [124]). It appears that this important information is often not given in application papers. Generally, the higher the voltage (power), the better the signal-to-noise ratio. Thus low-voltage excitation usually requires advanced signal processing procedures for damage detection. High-voltage excitation (i.e. voltage above/below $\pm$ 20V) require power amplification of high-frequency excitation signals. They also tend to increase the drift in the responses, significantly deteriorating the resolution capabilities of the data acquisition system [79]. High-power/voltage signals are not applicable when low power/voltage is required from the damage detection system or in hazardous conditions. Often Lamb wave responses are amplified before data acquisition and signal processing analysis (e.g. [125])

4.5 Signal Features Related to Damage

Various features from Lamb wave responses are used for damage detection. The analysis utilises either direct responses and/or reflected signals. Scattering, attenuation and/or mode conversion of Lamb waves are the most commonly used features for damage detection. However, wave scattering may be caused not only by material and structural defects but also by various structural boundaries such as stiffeners, joints or rivets. The attenuation mechanism of Lamb waves is very complex and includes the reduction in signal amplitude due to [77]: dispersion, transducer beam spreading, energy conversion into heat (material damping), acoustic energy radiation into the surrounding media and scattering. Only the last element is used in practice for damage detection. However, the amplitude of Lamb wave responses can be affected by various structural boundaries and/or environmental conditions such as a change of temperature [126, 127]. The ratio between maximum amplitudes of various propagating modes seems to be more efficient that the actual maximum Lamb wave response amplitude [128].

The studies in [129, 130, 131] show that often mode conversion appears in Lamb wave responses due to damage in composite and metallic plates. The positions of local minima in the envelope functions of Lamb wave responses can indicate the presence of damage.

It is well known that the velocity of wave propagation can change due to damage, as explained in Section 4.4. As a result, Lamb wave responses are delayed. The time-of-flight is the most commonly used parameter utilising this information, as illustrated in [115, 132, 133, 134]. Often frequency/phase variations are also observed. However, the frequency content of Lamb wave responses is not widely analysed; application examples include [11, 109, 118, 125].

5 Modelling and Signal Processing

Modelling, signal processing and computation are crucial elements in the implementation and operation of any damage detection system. Some aspects related to guided ultrasonic waves are briefly discussed in this section. There are two important aspects related to modelling. These are: transducer design and wave propagation studies.

5.1 Transducer Modelling

The design of transducers is important for effective generation and sensing of guided ultrasonic waves. Modelling and simulation techniques are used to optimise the performance of transducers and to evaluate new transducers for damage detection applications. They are also performed to understand the physical mechanisms of ultrasonic wave propagation in transducer materials and energy transfer from transducers to monitored structures, as shown in [82]

Various modelling techniques are used for effective design and modelling of transducers. This includes three major approaches found in the literature: the equivalent circuit technique, Finite Element (FE) modelling and acoustic field analysis. The equivalent circuit models are widely used to predict the bandwidth of transducers. These models utilise capacitance, inductance and resistance elements in order to obtain the impedance which describes the mechanical radiation. The values of impedances and inductances can be estimated from the mechanical parameters of the transducer using various methods described in [135, 136], whereas the value of resistance can be obtained experimentally [137]. Applications examples of equivalent circuit models of ultrasonic transducers are given in [138, 139, 140, 141]. FE analysis models transducers as coupled oscillators using constitutive material equations. Piezoelectric materials are then divided into small elements for which mechanical displacements, forces, electrical charges and potentials are defined at their nodes. The analysis involves FE commercial tools well known in mechanical engineering (e.g. *ANSYS* with piezoelectric finite elements) [142], FE software for active material applications (e.g. *Atila* [143], *PZFlex* [144]), modelling software with multiphysics capabilities (e.g. *FEMLAB* [145]) and various programmes implemented in FORTRAN and C and supplemented by FE commercial packages (e.g. [146]). The FE modelling paradigm for ultrasonic transducers is well presented in [147], an excellent literature survey is given in [148, 149], whereas good application examples are given in [150, 151, 152, 153]. The acoustic field analysis leads to the description of transducer pressure variations in the ultrasonic field. A number of analytical approaches mostly based on Huygens's principle are used to obtain presure variations. This includes: Green's functions, Kirchoff's analysis, Rayleigh's approach and z-transforms, as shown in [1, 147, 154, 155, 156]. Axial pressure profiles and polar coordinate diffraction-type graphs are the most widely used methods for the presentation of pressure distributions, as described in [1, 15].

Although, all three approaches described above are important for transducer modelling, it appears that FE analysis is being adopted as the most practical technique for damage detection applications based on piezoceramics. Figure 7 gives an example of three different models of a piezoceramic transducer.

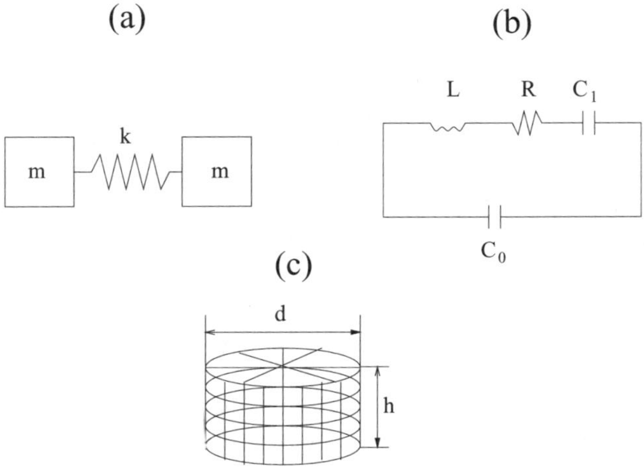

Fig. 7. Examples of piezoceramic transducer models: (*a*) harmonic oscillator (*b*) Van Dyke equivalent circuit model (*c*) FE model.

The are two major areas of interest related to transducer developments. The first area is related to acoustic field distribution. It is important to establish parameters related to the ultrasonic field such as directivity, near-field and angle of divergence of transducers, as described in [1]. The second area is the analysis of acoustic and vibration properties of transducers. These properties determine temporal behaviour and frequency characteristics of transducers. The temporal behaviour is usually studied when a transient response is required or non-stationary/non-linear phenomena are involved. The analysis often utilises the Laplace transform, z-transform and various discrete time-domain tools. Eigenvalue analysis, FRFs and transfer functions are used in order to obtain the frequency characteristics of transducers. The major interest is in the most effective bandwidths regarding the electro-mechanical energy transfer. Resonance frequencies of piezoceramic transducers are often used in practice as excitation frequencies in damage detection based on guided ultrasonic waves, as explained in Section 4.4. Acoustic and vibration properties of transducers can be also obtained using experimental procedures, as demonstrated in [157].

Piezoelectric materials exhibit coupled mechanical and electrical properties, as described in 3.1. The key element in all techniques used for transducer modelling is the way these electromechanical properties are described. The most common approach is to use the constitutive equations given as

$$\begin{aligned} S &= S_{\mathrm{e}}T + d^{t}E \\ D &= \epsilon_{T} + dT \end{aligned} \tag{13}$$

where S is the mechanical strain tensor, E is the electric field tensor, T is the mechanical stress tensor, D is the electrical displacement tensor, d is the piezoelectric coefficient vector, ϵ is the permittivity vector and t indicates the vector transpose. Subscripts E and T indicate that the appropriate quantities are under constant (usually zero) electrical and stress fields, respectively. Equation (14) gives the strain-charge form of the constitutive equations which relate the electric and mechanical fields. The other three forms give rearranged relationships between the state variables. There are two difficulties with modelling of the constitutive equations. Firstly, the voltage-strain behaviour of piezoceramics is in general non-linear. Secondly, most piezoceramics show a typical hysteretic behaviour which exhibits energy losses. Often electrical impedance is used in practice to describe dynamic properties of piezoceramic transducers [158, 159]. This function can be obtained when one of the electrodes is excited by the electrical charge and at the same time the other electrode is grounded. Resonance frequencies f_R (natural frequencies for short-circuited electrodes; electrical potential excitation applied) and anti-resonance frequencies f_A (natural frequencies for open-circuited electrodes; electrical charge excitation applied) are the most characteristic points on the impedance curves. The maximum response of the transducer is achieved when the excitation frequency is between f_R and f_A. For thickness mode vibration, the electromechanical coupling k can be determined from these frequencies using the following equation

$$k^2 = \frac{f_A^2 - f_R^2}{f_R^2} \qquad (14)$$

Alternatively, mechanical displacements are used to describe dynamic properties of transducers [160]. For narrow piezoelectric elements (i.e. w/t less than 1; w-width, t-thickness) the thickness mode always corresponds to the first natural frequency whereas for wide piezoelectric elements (i.e. w/t greater than 1) the thickness mode always corresponds to the second or higher natural frequencies [150]. The best dynamic performance is achieved when the excitation frequency is chosen as one of the natural resonance frequencies of the transducer. This enables a very efficient conversion from the electrical to mechanical energy, as discussed in Section 4.4. Resonance frequencies are associated with different vibration modes which in turn can be used for the generation of different types of Lamb waves [15, 161]. The thickness extensional or radial (for a disc-shaped transducer) mode of the piezoceramic generates the S_0 mode which is dominated by a longitudinal wave travelling in the plate. The S_1 and S_2 modes result from thickness shear modes of vibration whereas the S_3 mode can be generated by the thickness extensional mode. Also, the A_0 mode originates from the thickness extensional mode.

5.2 Wave Propagation Modelling

Knowledge and physical understanding of wave propagation in monitored structures can significantly ease the interpretation of damage detection results. The need for boundary conditions for the existence of Lamb waves is the major difficulty associated with analytical solutions. Thus modelling procedures are often used. A number of different numerical techniques have been developed for the analysis of wave propagation and modelling of wave interactions with damage.

The finite difference (FD) equations [162] and FE analysis [163] have been applied for wave propagation modelling for many decades. Application examples include [61, 109, 164, 165, 166, 167]. The boundary element method (BEM) [168] is similar to FE analysis. The major difference is that the BEM converts volume integrals to surface integrals using a fundamental solution in terms of special Green's functions. Application examples of wave propagation include [169, 170, 171]. Although, FE analysis is particularly well suited to problems of very complex geometries, BEMs are computationally more efficient for guided wave scattering in infinite media and less convenient for finite media.

Wave propagation modelling procedures can be developed from matrix formulations of wave equations. This leads to a matrix description of the analysed system, given in terms of its external boundaries. The solutions can be obtained from either responses or modal models. The former leads to amplitudes and phases of reflected and transmitted waves whereas the latter gives velocities and frequencies of propagating waves. A good summary and application examples of matrix techniques for modelling ultrasonic waves is given in [172]. The global matrix approach has also been used for wave propagation in [173, 174].

Both the FE analysis and BEMs are closely related to FD equations. FD equations utilise second- or higher-order approximations to space derivatives. FD-based methods are computationally expensive for large models and complex geometries. Much less storage data, due to lower level of discretisation and polynomial approximation, is required by the finite strip elements (FSE) approach [175, 176]. An overview of FD-based methods for wave propagation is given in [177].

FD methods approximate function derivatives using local variables on a small grid spacing. In contrast spectral element (SE) methods [178], which have emerged in the last twenty years, are global. SE methods approximate functions as a sum of smooth basis functions, e.g. trigonometric functions or Chebyshev polynomials, and then differentiate these functions exactly. These methods are easy to implement and efficient for wave propagation, as shown in [179, 180, 82]. The mass-spring lattice model (MSLM) has also been developed and applied for wave propagation in [181, 182]. The MSLM models the inertia of the analysed medium using lumped mass particles; the stiffness is modeled using spring elements connecting these particles. This model is in fact similar to the FD method since finite difference equations are used to describe spring constants. Nevertheless, the MSLM does not require any special treatment in the traction-free conditions for Lamb waves; disconnecting the springs is sufficient to fulfill these conditions [182].

FD-based methods are very convenient for wave propagation in media with homogeneous or continuously varying physical properties. However, boundaries and discontinuities between different types of media lead to approximate solutions and can produce severe errors [183]. More recently, the new Local Interaction Simulation Approach (LISA) [183, 184, 185] has been proposed to avoid these difficulties. Although the method is formally similar to the FD approach it does not use directly any finite difference equations but simulates wave propagation heuristically, i.e. directly from physical phenomena and properties. The LISA uses the so-called sharp interface model (SIM) imposing the continuity, or in other words perfect contacts, of displacement and stresses at interfaces and discontinuities. The LISA is computationally efficient and accurate to model complex wave propagation and diffusion problems in complex media [183, 184, 185, 186]. LISA applications for Lamb wave propagation include examples in metallic [121, 187] and composite [117, 118] structures. A novel approach with the LISA framework is the application of the spring model which substitutes the problem of wave propagation with the vibration problem of tensorial springs, as presented in [181].

In contrast to elastic wave propagation studies, wave interactions with damage has found relatively less attention in the literature. Most applications are related to wave scattering problems using classical, analytical and numerical, FD/FE based approaches [188]. The scattering theory for Lamb waves is not well developed. Application examples in this area include wave scattering from: thin slots, cracks or surface defects using the BE analysis [171, 189, 190, 191], finite cracks using the Wiener-Hopf method [192], circular inclusions using analytical Mindlin's higher-order plate theory [193], partially debonded fibres in fibre-reinforced composites using the null field theory [194, 195] and partially debonded piezoelectric fibres in composite plates using the dynamic theory of piezoelectricity [196]. Most of the scattering-related work in this area is based on the null-field approach in which the actuator/sensor configuration is used in an oblique rather than normal incident inspection.

Lamb wave interactions with damage in metallic and composite materials have also found some attention in the literature. Elastic wave interactions with various crack configurations in metallic structures using the MSLM approach has been used in [182]. Lamb wave examples in metallic structures include interactions with: a hole in plates using the FE analysis [197], surface defects in plates using a combined FE and modal expansion method [198], damage slots in circular pipes using the FE analysis [199] and damage slots in aluminium plates using LISA [187, 200, 201]. Research studies related to composite materials include Lamb wave interactions with: delaminations in composite plates using the FE analysis [119, 112] and and a hole in a composite plate using LISA [117, 118]. All these studies confirm that propagation of Lamb waves and wave interactions with damage are complex problems. Figure 8 gives an example of Lamb wave simulation in the aluminium plate. The in-plane and out-of-plane snapshots of the S_0 mode are presented. The wave propagates from the piezoceramic transducer located in the bottom of the plate. After around 22 μs the wave reaches the rectangular damage slot where clear scattering and then reflections from the top of the plate can be observed.

Simulations of Lamb wave interactions with damage require models of damage. Various techniques have been used in this area including continuous models (e.g. beam with a transverse crack or beam with delamination), discrete-continuous models (e.g. spring elements as cracks), FE techniques (e.g. separation of nodes, mesh condensation, introduction of mid-side nodes, special elements, extra boundary conditions), BE techniques, transition matrix methods, graph methods and analogous techniques, as summarised in [202].

5.3 Signal Processing for Damage Detection

Lamb wave based damage detection system requires the availability of appropriate signal processing to analyse features from sensor data and to translate this information into a diagnosis of location and severity of damage. The overall chain of processing includes pre-processing techniques, feature extraction/selection, pattern recognition, sensor location techniques, sensor validation methods and data fusion, as illustrated in [203, 80]. An overview of various signal processing techniques for SHM and Smart Structures/Materials applications is given in [204] and [108], respectively. A vast amount of literature exist on the subject. Figure 9 shows a selection of advanced signal processing techniques with application examples to guided ultrasonic waves.

Once the optimal transducer scheme (see Section 3) and monitoring strategy (see Section 4) are chosen, signal processing is the crucial element for effective damage detection. The major problem is how to extract/select reliable signal features which are sensitive to various types of damage yet insensitive to any other effects, e.g. external loads, operational or environmental conditions. Although a number of signal processing studies have been performed to establish Lamb wave features related to damage, only a few attempts have been made to analyse features insensitive to undesired effects. This includes studies of temperature effects on damage detection [126, 127].

All measurements are subject to uncertainty. Lamb wave measurement uncertainties are associated with experimental data acquisition, environmental conditions or damage evolution. Failure to build uncertainties may result in false damage alarms, costly inspections or potential catastrophes when damage is not detected. Various modelling techniques have been developed to

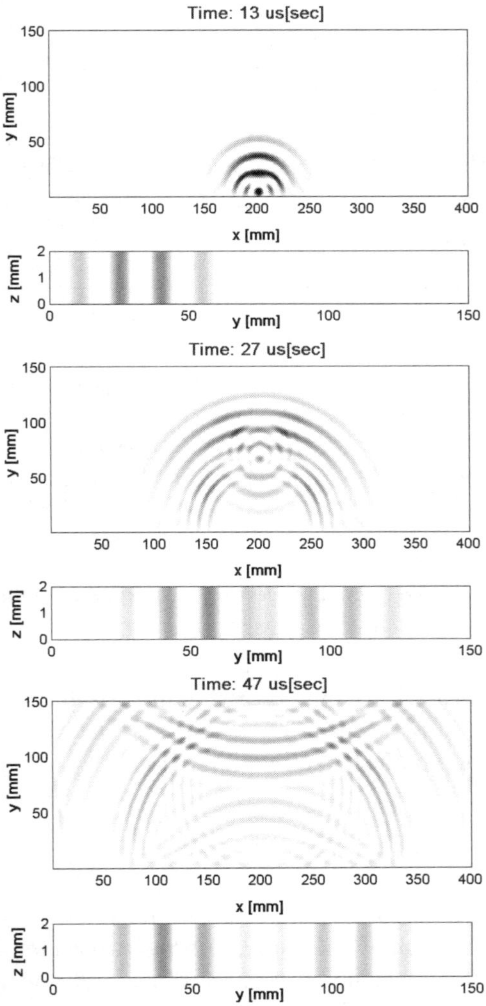

Fig. 8. LISA model of the S_0 Lamb wave mode propagating out-of-plane (top figures) and in-plane (bottom figures) in the aluminium plate with a defect. Numerical simulations after: (*a*) 13 μs (*b*) 27 μs (*c*) 47 μs [186].

analyse uncertainties. This includes: probabilistic, fuzzy-set and convex models. Very little research related to Lamb wave inspection has been done in this area. Application examples include [205, 206].

Finally, once damage is detected, located and assessed, a further analysis is to predict the remaining service life of the structure and possible breakdown. This activity, often called damage prognosis, falls into the combined field of fatigue analysis, fracture mechanics, design assessment, pattern recognition for damage detection, statistical analysis, novel model validation and

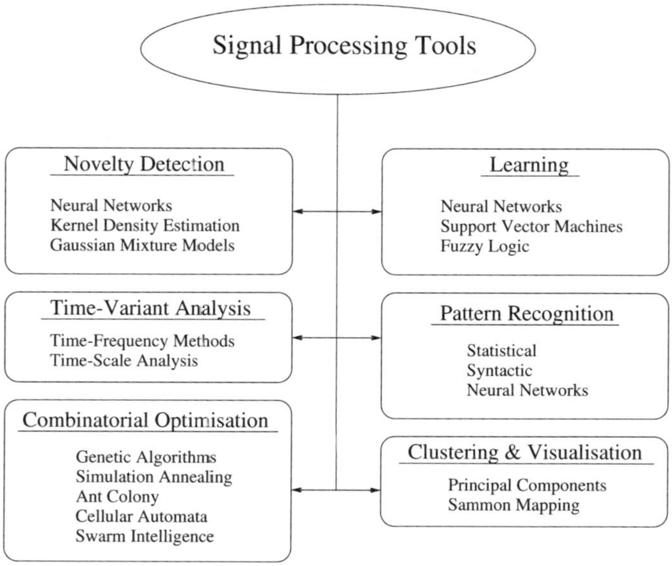

Fig. 9. Selection of advanced signal processing techniques used for damage detection applications.

uncertainity quantification techniques, reliability-based decision making algorithms, as described in [207, 208, 209]. It appears that very little research has been performed and experience gathered in this area.

6 Application Examples

Guided ultrasonic waves have found many damage detection applications in various areas of engineering. Papers on Lamb waves used for damage detection are scattered in different conference proceedings and journals published by various engineering communities such as NDT/NDE, SHM, physical acoustics and smart materials/structures. It is difficult to count and report the number of papers published every year. Very few publications cover practical aspects related to design, implementation testing and validation of SHM systems based on guided ultrasonic waves. One of the first attempts is given in [210], where all these aspects are discussed in the context of aerospace engineering.

Although the majority of Lamb wave publications are related to metallic and composite structures, some attempts have been made to apply guided ultrasonic waves for damage detection in concrete and even wooden structures. An excellent selection of papers from the NDT/NDE community is presented in [27, 211].

Tables 2 and 3 give only application examples of publications on guided ultrasonic waves for structural damage detection in various areas of engineering.

In what follows, two case studies from the metallic and composite structures are also presented.

Table 2. Application examples of guided ultrasonic waves for structural damage detection in metallics.

Material	Component	Damage	References
aluminium	plate	hole	[72, 73, 75]
		notch	[131, 212, 213]
		thinning	[214]
		corrosion	[24]
		fatigue crack	[11, 128, 210, 215]
			[216, 72, 128, 187]
		stress-corrosion crack	[217]
		quality of CFRP repair	[218, 219]
	pipe	notch	[220]
	automotive panel	weld defect	[221]
	bonded plates	ageing bond	[58]
		disbond	[222]
	riveted panel	fatigue crack	[42, 125, 210, 223]
	tear strap	corrosion	[24]
	honeycomb (aluminium skin and core)	delamination	[224]
	lap joint	disbond	[25, 132]
		corrosion	[24]
		environmental degradation	[225]
	aircraft lap joint	fatigue crack	[116, 122, 210, 226]
	aircraft fuselage	crack	[210, 227, 228]
	aircraft panel	corrosion	[229]
steel	plate	weld defect	[109]
		thinning	[230]
	pipe	saw cut	[231]
		corrosion	[232, 233, 234]
	cylinder	crack	[235]
	stud/bolt	deformation, breakage	[236, 237, 238]
	rail	crack	[238, 239, 240, 241]
		breakage	[242]
		weld defect	[238, 239, 240]
	bridge tendon	fracture	[243, 244]
		corrosion	[243, 244]
	ship hull	battle damage	[245, 246]
cast iron	pipe	crack	[247]
titanium	bonded plates	disbond	[248, 249]

Table 3. Application examples of guided ultrasonic waves for structural damage detection in materials other than metallics.

Material	Component	Damage	References
plexiglass	plate	hole	[40, 128]
plastics	pipe	crack	[250]
		weld defect	[250]
graphite/epoxy	plate	fatigue damage	[111, 251]
		simulated delamination	[76]
	stiffened plate	disbond	[8]
	filament wound tube	impact damage	[133]
	sandwiched cylinder	delamination	[8]
CF pre-preg	plate	simulated delamination	[129]
CFRP	beam	simulated delamination	[79]
	plate	simulated delamination	[112, 252, 253]
		impact damage	[31, 110, 210]
		fatigue damage	[111]
		thermal damage	[111]
	stiffened plate	disbond	[210]
	honeycomb (composite skin -aluminium core)	delamination	[224, 254]
	wing-box	impact damage	[210]
	aircraft rudder	delamination	[254]
GRP	plate	impact damage	[31, 110]
concrete	block	discontinuity	[255]
		surface crack	[241]
		quality of CFRP repair	[256, 257]
composite	thick multilayered panel (ceramic tiles, glass/epoxy,rubber)	impact damage	[258]
reinforced concrete	block	steel/concrete interface	[259, 260]
wood	beam	decay, knots, voids	[261]
	plate	decay, knots, voids	[261]
	plywood plate	disbond	[262]

A problem currently studied in [72, 74, 75] is related to laser vibrometry for damage detection in metallic structures. The A_0 Lamb wave mode was generated using a piezoceramic actuator bonded on the aluminium plate. The Lamb wave response data were acquired at various locations on the plate using a scanning laser vibrometer. The top part of Figure 10 shows the location of the actuator, damage slot and monitored area. The damage is clearly exhibited by the locally increased amplitude in the bottom part of Figure 10. The method allows for rapid and simple monitoring of large structures.

Composite materials have become widely used in many areas of engineering applications. Many inspection problems with composites in the aircraft industry are related to detection and location of impact damage. A simple drop-weight experiment, using the equipment shown in the top part of Figure 11, was performed in [263] to study impact damage detection in a composite plate. Low profile piezoceramic transducers were bonded on the monitored structure in order to propagate and acquire Lamb waves. The bottom part of Figure 11 presents an example of the Lamb wave response data together with the wavelet transform plot. Advanced signal processing can be used to identify damage features related to various types of damage such as delamination or fibre/matrix cracking.

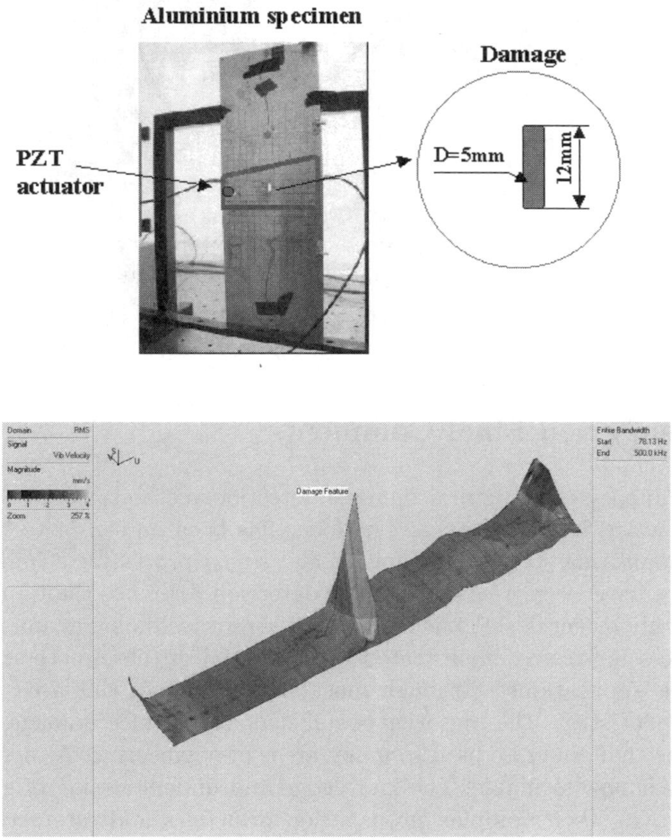

Fig. 10. Damage detection in a metallic specimen using Lamb waves and a scanning laser vibrometer. The aluminium plate is shown in the top figure whereas damage detection results are given in the bottom figure.

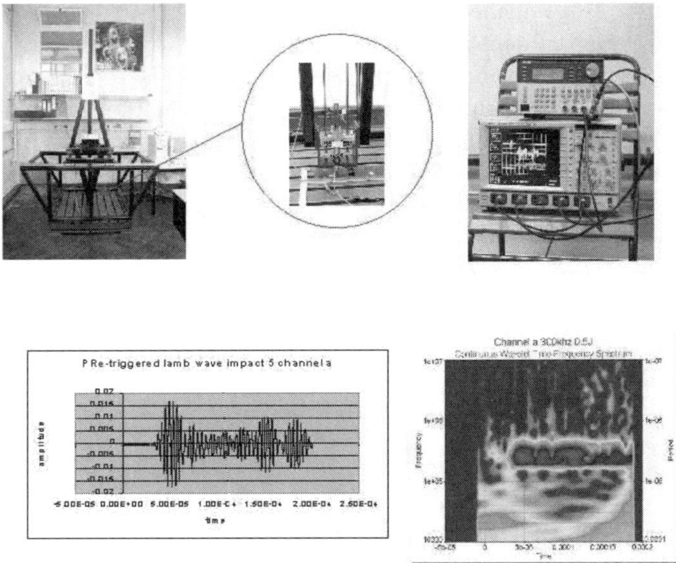

Fig. 11. Damage detection using Lamb waves and integrated piezoceramic transducers. Experimental arrangements are shown in the top figure whereas the Lamb wave data together with the wavelet transform plot is presented in the bottom figure.

7 Summary and Final Comments

A brief summary of structural damage detection methods based on guided ultrasonic waves has been given. The focus has been on sensor technologies, damage monitoring strategy, modelling and signal processing. Guided ultrasonic waves have been used for damage detection for more than forty years. However, only recent developments in smart sensor technologies and advanced signal processing have demonstrated the potential of these methods for real engineering applications. Although most of the work in this area is still in a development stage, the maturity is sufficient to produce damage monitoring systems that could be used in many areas of engineering. Altogether, the technology is now available. The knowledge and understanding of guided ultrasonic waves, their complex propagation principles and interactions with defects is now better understood then 10-20 years ago. A large selection of different types of transducers based on smart sensor technologies is available. Reliable signal processing techniques have been developed for damage detection. The question is thus what needs to be done to transfer these interesting academic and laboratory developments to real engineering applications. An integration of appropriate technologies and a large testing and validation effort is required in order to develop reliable SHM systems. It is clear that often new inspection/maintenance and design philosophies will be needed to achieve this. However, it is now inevitable that systems based on guided ultrasonic waves will be one of the core technologies available for damage inspection and monitoring of structures in the near future. The attractive potential of such systems is related to life-cycle cost and safety aspects of structures.

Acknowledgement. I would like to pass my thanks to Boon Lee and Daniel Betz, my superior PhD students, for the fantastic work, inspiring discussions and motivation. My acknowledgements also go to Laurent Mallet and Boon Lee for their contributions related to the literature survey in Sections 3.1 and 5.2, respectively. The dispersion curve presented in Figure 3 have been computed by Boon Lee. The top part of Figure 10 and Figure 11 have been provided by my students Laurent Mallet and Kshitij Singh, respectively. The laser data in Section 6 have been gathered with the help of Dr Roger Trayner from *Polytec, UK*. Finally, I would like to thank Prof. Keith Worden and Dr Gareth Pierce for their valuable comments on the material presented in this chapter.

References

1. Rose J L (1999) Ultrasonic waves in solid media. Cambridge University Press, Cambridge
2. Scott IG (1991) Basic acoustic emission. Gordon and Breach Publishers, New York
3. Muravin G (2000) Inspection, diagnostics and monitoring of construction materials and structures by the acoustic emission methods. Minerva Press, London
4. Holroyd T (2001) The Acoustic emission and ultrasonic monitoring handbook. Coxmoor Publishing Company, Oxford
5. Buirks AS, Green Jr. RE, McIntire P (eds) (1991) Nondestructive Testing Handbook. Vol. 7, Ultrasonic Testing. American Society of Nondestructive Testing Press, Columbus
6. Worlton DC (1957) Ultrasonic testing with Lamb waves. Nondestructive Testing 15(4): 218-222
7. Anderson EH (1989) Piezoceramic induced strain actuation of one and two dimensional structures. PhD Thesis. Massachusetts Institute of Technology
8. Kessler SS, Spearing SM (2001) Damage detection in built-up composite structures using Lamb wave methods. Submitted to Journal of Intelligent Materials, Systems and Structures
9. Vary A (1982) Acousto-ultrasonic characterisation of fiber reinforced composites. Materials Evaluation 40: 650-654
10. Vary A (1988) The Acousto-ultrasonic approach. In: Duke Jr. J C (ed) Acousto-Ultrasonics: Theory and Application. Plenum Press, New York
11. Biemans C, Staszewski WJ, Boller C, Tomlinson GR (2001) Crack detection in metallic structures using broadband excitation of acousto-ultrasonics. Journal of Intelligent Materials Systems and Structures 12(8): 589-597

12. Lamb H (1917) On waves in an elastic plate. Proceedings of the Royal Society 93 (PT series A): 114-128
13. Viktorov IA (1967) Rayleigh and Lamb waves - physical theory and applications. Plenum Press, New York
14. Achenbach JD (1984) Wave propagation in elastic solids. North-Holland, New York
15. Auld BA (1990) Acoustic fields and waves in solids. 2nd edition. Kreiger, Malabar
16. Pavlakovic BN, Lowe MJS, Alleyne DN, Cawley P (1997) Disperse: a general purpose program for creating dispersion curves. In: Thompson DO, Chimenti DE (eds) Annual Review of Progress in Quantitative NDE, Vol. 16. Plenum Press, New York, 185-192
17. Web page: `www.imperial.ac.uk/research/ndt/public/productservice/disperse.htm`
18. Wilcox P, Evans M, Diligent O, Lowe MJS, Cawley P (2002) Dispersion and excitability of guided acoustic waves in isotropic beams with arbitrary geometries. In: Thompson DO, Chimenti DE (eds) Review of Progress in Quantitative NDE, American Institute of Physics 21: 203-210
19. Trainer M (2003) Kelvin and piezoelectricity. European Journal of Physics 24: 535-542
20. Web page: `www.morganelectroceramics.com/piezoguide1.html`
21. Web page: `www.piezo.com`
22. Nayfeh A, Chimenti DE (1988) Propagation of guided waves in fluid-coupled plates of fiber-reinforced composites. Journal of the Acoustical Society of America 83: 1738-1743
23. Krautkramer J, Krautkramer H (1990) Ultrasonic testing of materials. Springer Verlag, Berlin
24. Chahbaz A, Mustapha V, Hay B (1996) Corrosion detection in aircraft structure using guided Lamb waves. The e-Journal of Nondestructive Testing 1(11): `www.ndt.net/article/tektrend/tektrend.htm`
25. Chahbaz A, Mustapha V, Hay B, Brassard M, Dubois S (1997) Corrosion detection in aircraft structure using guided Lamb waves The e-Journal of Nondestructive Testing 2(3): `www.ndt.net/article/tektren2/tektren2.htm`
26. Zhenqing L (2000) Lamb wave analysis of acousto-ultrasonic signals in plate. In: Proceedings of the 15th World Congress on Nondestructive Testing, Rome, Italy. Paper No. 602: `www.ndt.net/article/wcndt00/papers/idn602/idn602.htm`
27. Rose JL (2002) A baseline and vision of ultrasonic guided wave inspection potential. Journal of Pressure Vessel Technology 124: 273-282
28. Egusa S and Iwasawa N (1993) Polling characteristics of PZT/epoxy piezoelectric paint. Ferroelectrics 145: 45-60
29. Chang FK (1998) Smart layer: built-in diagnostic system for composite structures. In: Proceedings of the Fourth European Conference on Smart Structures and Materials, Second International Conference on on Micromechanics, Intelligent Materials and Robotics, Harrogate, UK. 777-786
30. Lin M, Powers WT, Qing X, Kumar A, Beard JS (2002) Hybrid piezoelectric/fibre optic SMART layers for structural health monitoring. In: Proceedings of the First European Workshop on Structural Health Monitoring, Paris, France. 641-648

31. Badcock RA, Birt EA (2000) The use of 0-3 piezocomposite embedded Lamb wave sensors for detection of damage in advanced fibre composites. Smart Materials and Structures 9: 291-297
32. Huynen HHJ, Dijkstra FH, Bouma T, Wustenberg H (2000) Advances in ultrasonic transducer development. In: Proceedings of the 15th World Congress on NDT, Rome, Italy. Paper No. 48: `www.ndt.net/article/wcndt00/ind048/ind048.htm`
33. Web page: `www.advancedcerametrics.com`
34. Web page: `www.acellent.com`
35. Aliouane S, Hassam M, Badidi Bouda A, Benchaala A (2000) Electromagnetic Acoustic Transducers (EMATs) design evaluation of their performances. In: Proceedings of the 15th World Congress on Nondestructive Testing, Rome, Italy. Paper No. 591: `www.ndt.net/article/wcndt00/papers/idn591/idn591.htm`
36. Li J, Rose JL (2001) Guided wave inspection of containment structures. Materials Evaluation 59: 783-787
37. Ditri JJ, Rose JL, Pilarski A (1993) Generation of guided waves in hollow cylinders by wedge and comb type transducers. Review of Progress in Quantitative Nondestructive Evaluation 12A: 211-218
38. Wilcox P, Castaings M, Monkhouse R, Cawley P, Love M (1997) An example of the use of interdigital PVDF transducers to generate and receive a high order Lamb wave mode in a pipe. Review of Progress in Quantitative Nondestructive Evaluation 16: 919-26.
39. Monkhouse RSC, Wilcox PD, Cawley P (1997) Flexible integrated PVDF transducers for the generation of Lamb wave in structure. Ultrasonics 35: 489-498
40. Blanquet P, Demol T, Delebarre C (1996) Application of array transducers to health monitoring of aeronautic structures. In: Proceedings of the 14th World Congress on Nondestructive Testing, New Delhi, India. 4: 2057-2060
41. Pouget J, Marguet J, Pichonnat F, Chupin L (2002) Phased array technology: concept, probes and applications. The e-Journal of Nondestructive Testing 7(5): `www.ndt.net/article/v07n05/pouget/pouget.htm`
42. Kress KP, Dittrich K, Guse G (2001) Smart wide-area imaging sensor system (SWISS). In: Proceedings of the SPIE International Symposium on Smart Structures and Materials, Conference on Industrial and Commercial Applications of Smart Structures Technologies, San Diego, California. Paper No. 4332-62
43. Ritter J (1996) Universal phased array UT probe for nondestructive examinations using composite crystal technology. The e-Journal of Nondestructive Testing 1(12): `www.ndt.net/article/ritter/ritter.htm`
44. Khuri-Yakub BT, Cheng CH, Degertekin FL, Ergun S, Hansen S, Jin XC, Orlakan O (2000) Silicon micromachined ultrasonic transducers. Japanese Journal of Applied Physics 39: 2882-2887
45. Yaralioglu GG, Badi MH, Ergun AS, Cheng AS, Khuri-Yakub BT (2001) Lamb wave devices using micromachined ultrasonic transducers Applied Physics Letters 78(1): 111-113
46. Betz D, Thursby G, Culshaw B, Staszewski WJ 2003 Acousto-ultrasonic sensing using fiber Bragg gratings. Smart Materials and Structures 12: 122-128
47. Betz D, Thursby G, Culshaw B, Staszewski WJ (2003) lamb wave detection and source location using fiber Bragg grating rosettes. In: Proceedings of the

10th International Symposium on Smart Structures and Materials, Conference on Smart Sensor technology and Measurement Systems. Paper No. 5050-18
48. Hutchins DA, Wright WMD, Hayward G, Gachagan A (1994) Air-coupled piezoelectric detection of laser-generated ultrasound. IEEE Trans. Ultrason. Ferroelec. Freq. Contr. 41: 796-805
49. Castaings M, Cawley P (1996) The generation, propagation and detection of Lamb waves in plates using air-coupled ultrasonic transducers. Journal of the Acoustical Society of America 100(5): 3070-3078
50. Sorazu B, Thursby G, Culshaw B, Dong F, Pierce SG, Yong Y, Betz D (2003) Optical generation and detection of ultrasound. Strain 39(3): 111-114
51. Scruby CB, Drain LE (1990) Laser ultrasonics. Adam Hilger, New York
52. Hutchins DA (1986) Mechanisms of pulsed photoacoustic generation. Canadian Journal of Physics 64: 1247-1264
53. Pierce SG, Culshaw B (1998) Laser-generation of ultrasonic Lamb waves using low power optical sources. IEE Proceedings Science, Measurement & Technology 145(5): 244-249
54. Sontag H, Tam AC (1985) Optical monitoring of photoacoustic pulse propagation in silicon wafers. Applied Physics Letters 46(8): 725-727
55. Dewhurst RJ, Edwards C, Mckie AD, Palmer SB (1987) Estimation of the thickness of thin metal sheet using laser generated ultrasound. Applied Physics Letters 51(14): 1066-1068
56. Pierce SG, Culshaw B, Philp WR, Lecuyer F, Farlow R (1997) Broadband Lamb wave measurements in aluminium and carbon/glass fibre reinforced composite using noncontacting laser generation and detection. Ultrasonics 35: 105-114
57. Pierce SG, Culshaw B, Shan Q (1998) Laser-generation of ultrasound using a modulated continuous wave laser diode. Applied Physics Letters 72(9): 1030-1032
58. Heller K, Jacobs LJ, Qu J (2000) Characterization of adhesive bond properties using Lamb waves. NDT&E International 33(8): 555-563
59. Telschow KL, Bruneel SW, Walter JB, Koo LS (1997) Noncontacting ultrasonic and electromagnetic HTS tape NDE IEEE Trans. Appl. Superconductivity 7(2): 1319-1322
60. Ing RK, Monchalin JP (1991) Broadband optical detection of ultrasound by two-wave mixing in a photorefractive crystal. Applied Physics Letters 59(25): 3233-3235
61. Fromme P (2001) Defect detection in plates using guided waves. PhD Thesis No. 14397. Institute of Mechanical Systems. Swiss Federal Institute of Technology, Zurich
62. Atherton K, Dong F, Pierce G, Culshaw B (2000) Mach-Zehnder optical fibre interferometers for ultrasonic detection. In: Proceedings of the SPIE 7th International Symposium on Smart Structures and Materials, Conference on Sensory Phenomena and Measurement Instrumentation for Smart Structures and Materials, Newport Beach, California. Paper No. 3986-14 Paper No. 3986-04.
63. Monchalin JP (1986) Optical detection of ultrasound. IEEE Trans. UFFC 33: 485-499
64. Dewhurst RJ, Shan Q (1999) Optical remote measurement of ultrasound. Meas. Sci. Technol. 10: 139-168
65. Sriram P, Hanagud S, Craig JI (1992) Scanning laser Doppler techniques for vibration testing. Experimental Techniques 21-26

66. Stanbridge AB, Ewins DJ (1999) Modal testing using a scanning laser Doppler vibrometer. Mechanical Systems and Signal Processing 13(2): 255-270
67. Ewins DJ (2000) Modal testing: theory and practice. 2nd edition Research Studies Press, Taunton
68. Pai PF, Oh Y, Lee SY (2002) Detection of defects in circular plates using a scanning laser vibrometer. Structural Health Monitoring 1: 63-88
69. Waldron K, Ghoshal A, Schulz MJ, Sundaresan MJ, Ferguson F, Pai PF, Chung JH (2002) Damage detection using finite elements and laser operational deflection shapes. Finite Elements in Analysis and Design 38: 193-226
70. He L, Kobayashi S (2001) Determination of stress-acoustic coefficients of Rayleigh wave by use of laser Doppler velocimeter. JSME Int. Journal, Series A, 44(1): 17-22
71. Nishizawa O, Satoh T, Lei X (1998) Detection of shear wave in ultrasonic range by using Doppler vibrometer. Rev. Sci. Instr. 69: 2572-2573
72. Mallet L, Lee BC, Staszewski WJ, Scarpa F (2003) Damage detection in metallic structures using laser acousto-ultrasonics. In: Proceedings of the 4th International Workshop on Structural Health Monitoring, Stanford, California. 765-771
73. Kehlenbach M, Kohler B, Cao X, Hanselka H (2003) Numerical and experimental investigation of Lamb wave interaction with discontinuities. In: Proceedings of the 4th International Workshop on Structural Health Monitoring, Stanford, California. 421-429
74. Staszewski WJ, Lee BC, Mallet L, Scarpa F (2003) Structural health monitoring using scanning laser vibrometry, part I: Lamb wave sensing. Submitted to: Smart Materials and Structures
75. Mallet L, Lee BC, Staszewski WJ, Scarpa F (2003) Structural health monitoring using scanning laser vibrometry, Part II: Lamb waves for damage detection. Submitted to: Smart Materials and Structures
76. Degertekin FL, Khuri-Yakub BT (1997) Lamb wave excitation by hertzian contacts with applications in NDE. IEEE Transactions on Ultrasonics, Ferroelectrics, and Frequency Control 44(4): 769-778
77. Wilcox PD, Dalton RP, Lowe MJS, Cawley P (1999) Mode transducer selection for long range Lamb wave inspection. In: Proceedings of the Third International Workshop on Damage Assessment Using Advanced Signal Processing Procedures - DAMAS, Dublin, Ireland. Trans Tech Publications Ltd, Uetikon-Zuerich, 152-161
78. Pierce SG, Culshaw B, Manson G, Worden K, Staszewski WJ (2000) The application of ultrasonic Lamb wave techniques to the evaluation of advanced composite structures. In: Proceedings of the SPIE 7th International Symposium on Smart Structures and Materials, Conference on Sensory Phenomena and Measurement Instrumentation for Smart Structures and Materials, Newport Beach, California. Paper No. 3986-14
79. Kessler SS, Spearing SM, Soutis C (2001) Optimization of Lamb wave methods for damage detection in composite materials. In: Proceedings of the 3rd Workshop on Structural Health Monitoring, Stanford, California
80. Staszewski WJ, Boller C (2002) Acoustic wave propagation phenomena modelling and damage mechanisms in ageing aircraft. In: Proceedings of the Aircraft Integrated Monitoring Systems Conference - AIMS, Garmisch-Partenkirchen, Germany. 169-184

81. Kessler SS, Hastings A, Blair K (2003) Optimization of Lamb wave actuation/sensing patch for health monitoring of composite structures. In: Proceedings of the SPIE International Symposium on Smart Structures and Materials, Conference on Smart Structures and Integrated Systems. Paper No. 5056-80
82. Lee BC, Palacz M , Krawczuk M, Ostachowicz W, Staszewski WJ (2003) Wave propagation in a sensor/actuator diffusion bond model. Journal of Sound and Vibration. Accepted for publication.
83. Wilcox PD, Cawley P, Lowe MJS (1998) Acoustic fields from PVDF interdigital transducers. IEE Proceedings on Science, Measurement & Technology 145: 250-259
84. Varadan VK, Varadan VV (1999) Wireless remotely readable and programmable microsensors and MEMS for health monitoring of aircraft structures. In: Proceedings of the 2nd International Workshop on Structural Health Monitoring, Stanford, California. 96-105
85. Lee BC (2004) Modelling of Lamb wave interactions with defects in metallic plates. Department of Mechanical Engineering, Sheffield University
86. Staszewski WJ, Worden K (2001) Overview of optimal sensor location methods for damage detection. In: Proceedings of the SPIE's 8th International Symposium on Smart Structures and Materials, Conference on Modeling, Signal Processing and Control in Smart Structures, Newport Beach, California. Paper No. 4326-21
87. Goldberg DE (1999) Genetic algorithms in search, optimization, and machine learning. Adison-Weslay Publishing Co., Inc., Reading
88. Otten RHJM, Van Ginneken LPPP (1989) The annealing algorithm. Kluwer, Boston
89. Glover F (1989) Tabu search part 1. ORSA Journal of Computing 1: 190-216
90. Glover F (1990) Tabu search part 2. ORSA Journal of Computing 2: 4-32
91. De Werra D, Hertz A (1989) Tabu search techniques: a tutorial and application to neural networks. OR Spectrum 11: 131-14
92. Shannon CE, Weaver W (1949) The mathematical theory of communication. University of Illinois Press, Urbana
93. Worden K, Tomlinson GR (1993) Fault detection and sensor location using neural networks. Technical report. University of Manchester. Department of Mechanical Engineering
94. Worden K, Burrows AP, Tomlinson GR (1995) A combined neural and genetic approach to sensor placement. In: Proceedings of the 13th International Modal Analysis Conference, IMAC-95, Nashville, Tennessee.
95. Staszewski WJ, Worden K, Tomlinson GR (1996) Optimal sensor placement for neural network fault diagnosis. In: Proceedings of the ACEDC'96, Plymouth, UK.
96. Wong CS, Staszewski WJ (1998) Mutual information approach to sensor location for impact damage detection of composite structures. In: Proceedings of the International Conference on Noise and Vibration Engineering - ISMA 23, Leuven, Belgium. 1417-1422
97. Said WM, Staszewski WJ (2000) Optimal sensor location for damage detection using mutual information. In: Proceedings of the International Conference on Adaptive Structures and Technologies (ICAST), Nagoya, Japan. Technomic Publishing Co. Inc. 428-435

98. Raghavan H, Shah SL (2002) Diagnosis of sensor faults in closed-loop systems using models identified through subspace techniques. In: Proceedings of the Annual Meeting of the American Institute of Chemical Engineers, Indianapolis.
99. Chow EY, Willsky AS (1984) Analytical redundancy and the design of robust failure detection system. IEEE Transactions on Automatic Control AC-29: 603-614
100. Gertler J, Singer D (1990) A new structural framework for parity equation based failure detection and isolation. Automatica 26: 381-388
101. Frank PM (1994) On-line fault detection in uncertain nonlinear systems using diagnosis observers: a survey. Journal of Systems Science 25(12): 2129-2154
102. Isermann R (1984) Process fault detection based on modelling and estimation methods - a survey. Automatica 20: 387-404
103. Wald A (1945) Sequential tests of statistical hypotheses. Annals of Mathematical Statistics 16: 117-186
104. Upadhyaya BR, Eryurek E (1992) Application of neural networks for sensor validation and plant monitoring. Nuclear Technology 97: 170-176
105. Hines JW, Uhrig RE (1998) Use of autoassociative neural networks for signal validation. Journal of Intelligent and Robotic Systems 2: 143-154
106. Gribok AV, Urmanov AM, Hines JW, Uhrig RE (2003) Use of kernel based techniques for sensor validation in nuclear power plants. In: Bozdogan H (ed) Statistical data mining and knowledge discovery. Chapman & Hall/CRC Press, Boca Raton, Florida
107. Friswell MI, Inman DJ (1999) Sensor validation for smart structures. In: Proceedings of the 17th International Modal Analysis Conference, Orlando, Florida. 483-489
108. Worden K, Staszewski WJ (2003) Data fusion - the role of signal processing for smart structures and systems. In: Worden K, Bullough WA, Haywood J (eds) Smart Technologies. World Scientific, New Jersey
109. Alleyne DN, Cawley P (1992) Optimization of Lamb waves inspection techniques. NDT & E International 25(1): 11-22
110. Monnier T, Guy P, Jayet Y, Baboux JC (1999) Health monitoring of composites plates through Lamb wave analysis. Technical report INSA, Lyon: www.insa-lyon.fr
111. Seale MD, Smith BT, Prosser WH (1998) Lamb wave assessment of fatigue and thermal damage in composite. Journal of the Acoustical Society of America 103(5): 2416:2424
112. Guo N, Cawley P (1993) The interaction of Lamb waves with delaminations in composite plates. Journal of the Acoustical Society of America 94(4): 2240-2246
113. Lowe MJS, Challis RE, Chan CW (2000) The transmission of Lamb waves across adhesively bonded lap joints. Journal of the Acoustical Society of America 107(3): 1333:1345
114. Seifried R, Jacobs LJ, Qu J (2002) Propagation of guided waves in adhesive bonded components. NDT & E International 35: 317-328
115. Valle C, Niethammer M, Qu J, Jacobs LJ (2001) Cracks characterization using guided circumferential waves. Journal of the Acoustical Society of America 103: 2416-2424
116. Boller C, Ihn JB, Staszewski WJ, Speckmann H (2001) Design principle and inspection techniques for long life endurance of aircraft structures. In: Proceedings of the 3rd International Workshop on Structural Health Monitoring, Stanford, California: 275-283

117. Agostini V, Baboux JC, Delsanto PP, Monnier T, Olivero D (1999) Application of Lamb Waves for the Characterization of Composite Plates. In: Proceedings of the 9th International Symposium on Nondestructive Characterization of Materials, Sydney, Australia
118. Agostini V, Delsanto PP and Zoccolan D (2000) Flaw detection in composite plates by means of Lamb waves. In: Proceedings of the 15th World Congress of NDT, Rome, Italy. Paper No. 275: www.ndt.net/article/wcndt00/papers/idn275/idn275.htm
119. Alleyne DN, Cawley P (1992) The interaction of of Lamb waves with defects. Journal of the Acoustical Society of America 39: 381-397
120. Cawley P (1999) Quick Inspection of Large Structures Using Low Frequency Ultrasound. In: Proceedings of the 1st International Workshop on Structural Health Monitoring, Stanford, California. Technmic Publication: 529-540
121. Lee BC, Staszewski WJ (2003) Modelling of Lamb waves for damage detection in metallic structures. Part I: Wave propagation. Smart Structures and Materials : 12(5): 804-814
122. Ihn JB, Chang FK, Speckmann H (2001) Built-in diagnostics for monitoring crack growth in aircraft structures. In: Proceedings of the 4th International Conference on Damage Assessment of Structures (DAMAS), Cardiff, UK: 299-308
123. Monkhouse RSC, Wilcox PW, Lowe MJS, Dalton RP, Cawley P The rapid monitoring of structures using interdigital Lamb wave transducers Smart materials and Structures 9: 304-309
124. Buckley J (2000) Air-coupled ultrasound - a millennial review. In: Proceedings of the 15th World Congress on NDT, Rome, Italy. Paper No. 507: www.ndt.net/article/wcndd00/papers/idn507/idn507.htm
125. Grondel S, Moulin E, Delebarre C (1999) Lamb wave assessment of fatigue damage in aluminium plate. In: Proceedings of the 6th SPIE International Symposium on Smart Structures and Materials, Newport Beach, California
126. Manson G, Pierce SG, Worden K, Monnier T, Guy P, Atherton K (2000) Long-term stability of normal condition data for novelty detection. In: Proceedings of the 7th International Symposium on Smart Structures and Materials, Conference on Smart Structures and Integrated Systems. Paper No. 3985-34
127. Lee BC, Manson G, Staszewski WJ (2003) Environmental effects on Lamb wave responses from piezoceramic sensors. In: Proceedings of the 5th International Conference on Modern Practice in Stress and Vibration Analysis, Glasgow, Scotland: 195-202
128. Betz D (2004) Application of Optical Fibre Sensors for Structural Health Monitoring. PhD Thesis. Department of Mechanical Engineering. Sheffield University, UK
129. Guo N, Cawley P (1994) Lamb wave reflection for the quick non-destructive evaluation of large composite laminates. Materials Evaluation 52(3): 404-411
130. Lowe MJS, Alleyne DN, Cawley P (1998) The mode conversion of the guided wave by a part circumferential notch in a pipe. ASME Journal of Applied Mechanics 65:649-656
131. Shin HJ, Song SJ (2000) Observation of Lamb wave mode conversion on an aluminium plate. In: Proceedings of the World Congress on NDT, Rome, Italy. Paper No. 511: www.ndt.net/article/wcndt00/papers/idn511/idn511.htm

132. Sun KJ, Johnston PH (1993) Disbond detection in bonded aluminium joints using Lamb wave amplitude and time-of-flight. In: Proceedings of the 20th Review in Progress in Quantitative Nondestructive Evaluation, Brunswick, Maine
133. Beard S, Chang FK (1997) Active damage detection in filament wound composite tubes using built-in sensors and actuators. Journal of Intelligent Materials Systems and Structures 8: 891-897
134. Giurgiutiu V, Zagrai A, Bao JJ (2002) Embedded active sensors for in-situ structural health monitoring of thin-wall structures. Transactions of ASME, Journal of Pressure Vessel Technology 124: 293-302.
135. Hueter TF, Bolt RH (1955) Sonics. Wiley, New York
136. Stansfield D (1990) Underwater electroacoustic Transducers. Bath University Press and Institute of Acoustics
137. Terunuma K, Nishigaki S (1983) A method for measuring the equivalent circuit elements for a piezo-resonator. Japanese Journal of Applied Physics 22 suppl. 22-2 143-5
138. IEEE Standard on Piezoelectricity (1987) IEEE/ANSI 176
139. Ikeda T (1990) Fundamentals of piezoelectricity. Oxford University Press, Oxford
140. Sherrit S, Wiederic HD, Mugherjee BK, Sayer M (1997) An accurate equivalent cicruit for the unloded piezoelectric vibrator in the thickness mode. Journal of Applied Physics 30: 2354-bitemhayward:system
141. Hayward G, MacLeod, Durrani TS (1984) A system model of the thickness mode piezoelectric transducer. Journal of the Acoustical Society of America 76(2): 369-382
142. Web page: `www.ansys.com/ansys/multiphysics.htm`
143. Web page: `www.cedrat-grenoble.fr/software/atila/atila.htm`
144. Web page: `www.wai.com/AppliedScience/Software/Pzflex/index-pz.html`
145. Web page: `www.femlab.com`
146. Lerch R (1988) Finite element analysis of piezoelectric transducers. In: Proc. IEEE Ultrasonic Symp. 2: 643-654
147. Abboud NN, Wojcik GL, Vaughan DK, Mould J, Powell DJ, Nikodym L (1998) Finite element modeling for ultrasonic transducers. In: Proceedings of the SPIE International Symposium on Medical Imaging, San Diego, California
148. Kocbach J (1996) Finite element modelling of piezoelectric transducers. MSc Thesis (in Norwegian). Department of Physics. University of Bergen, Norway: `www.fi.uib.no/ jankoc/hfag/hoved.html`
149. Kocbach J (2000) Finite element modelling of piezoelectric transducers. PhD Thesis. Department of Physics. University of Bergen, Norway
150. Lerch R (1990) Simulation of piezoelectric devices by two- and three-dimensional finite elements. IEEE Transactions on Sonics and Ultrasonics 37: 233-247
151. Guo N, Cawley P (1992) The finite element analysis of the vibration characteristics of piezoelectric discs. Journal of Sound and Vibration 159(1): 115-138
152. Guo N, Cawley P (1992) Measurement and prediction of the frequency spectrum of piezoelectric disks by modal analysis. Journal of the Acoustical Society of America 92(6): 3379-3387
153. Lerch R, Kaltenbacher M, Landes H, Hoffelner J, Rausch M, Schinnerl M (2001) Advanced transducer modeling using the finite element method. In: Proceedings of the 11th International Symposium on Theoretical Electrical Engineering - ISTET218, Linz, Austria

154. Challis RE, Harrison JA (1983) Rapid solutions to the transient response of piezoelectric elements by z-transform techniques. Journal of the Acoustical Society of America 74(6): 1673-1680
155. Holland R (1966) Application of Green's functions and eigenmodes in the design of piezoelectric ceramic devices. PhD Thesis. Massachusetts Institute of Technology
156. Kuhnicke E, Volz U (1996) Designing ultrasonic transducers for layered media by means of Green's functions - a calculation software. In: Proceedings of the World Congress on NDT, New Delhi, India
157. Veidt M, Liu T, Kitipornchai S (2000) Experimental investigation of the acousto-ultrasonic transfer characteristics of adhesively bonded piezoceramic transducers. Smart Materials and Structures 9: 19-23
158. Hayward G (1984) A systems feedback representation of piezoelectric transducer operational impedance. Ultrasonics 22(4): 153-162
159. Wojcik GL, Vaughan DK, Abboud NN, Mould Jr. J (1993) Electromechanical modeling using explicit time-domain finite elements. In: IEEE Ultrasonic Symposium Proceedings 2: 1107-1112
160. Moulin E, Assaad J, Delebarre C, Kaczmarek H, Balageas D (1997) Piezoelectric transducer embedded in a composite plate: application to Lamb wave generation. Journal of Applied Physics 82(5): 2049-2055
161. Certon D, Felix N, Lacaze I, Teston F, Patat F (2001) Investigation of cross-coupling in 1-3 piezocomposite arrays. IEEE Transactions on Ultrasonics, Ferroelectrics and Frequency Control 48: 85-92
162. Strickwerda JC (1989) Finite difference schemes and partial differential equations. Wadsworth-Brooks, Belmont
163. Zienkiewicz OC (1989) The finite element method. Fourth edition. McGraw-Hill, London
164. Yamawaki H, Saito T (1992) Numerical calculation of surface waves using new nodal equation. Nondestructive Testing and Evaluation. 8-9: 379-389
165. Talbot RJ, Przemieniecki JS (1976) Finite element analysis of frequency spectra for elastic wave guides. International Journal of Solids Structures. 11: 115-138
166. Koshiba M, Karakida S, Suzuki M (1984) Finite element analysis of Lamb waves scattering in an elastic plate waveguide. IEEE Transactions on Sonics and Ultrasonics. 31(1): 18-25
167. Diaz S, Soutis C (2000) Structural integrity monitoring of CFRP laminates using piezoelectric devices. In: Proceedings of the 9th European Conference on Composite Materials, Brighton, UK
168. Brebbia CA, Tells JCF, Wrobel LC (1984) Boundary element techniques. Springer Verlag, Berlin
169. Cho Y, Rose JL (1996) A boundary element solution for mode conversion study of the edge reflection of Lamb waves. Journal of the Acoustical Society of America. 99(4): 2097-2109
170. Cho Y, Hongerholt DD, Rose JL (1997) Lamb wave scattering analysis for reflector characterization. IEEE Transactions on Ultrasonics, Ferroelectrics and Frequency Control 44: 44-52
171. Zhao X, Rose JL (2003) Boundary element modeling for defect characterzation potential in a wave guide. International Journal of Solids and Structures 40: 2645-2658

172. Lowe MJS (1995) Matrix techniques for modelling ultrasonic waves in multilayered media. IEEE Transactions on Ultrasonic, Ferroelectrics and Frequency Control. 42(4): 525-542
173. Lowe MJS (1993) Plate waves for the NDT of diffusion bonded titanium. PhD Thesis, University of London.
174. Lowe MJS, Cawley P (1994) The applicability of plate wave techniques for the inspection of adhesive and diffusion bonded joints. Journal of Nondestructive Evaluation. 13(4): 185-200
175. Cheung YK (1976) Finite strip method in structural analysis. Pergamon Press, Oxford
176. Liu GR, Tani J, Watanabe K, Ohyoshi T (1990) Harmonic wave propagation in anisotropic laminated strips. Journal of Sound and Vibration. 139(2): 313-330
177. Bond LJ (1990) Numerical techniques and their use to study wave propagation and scattering: a review. In: Datta SK, Achenbach JD, Rajapakse Y (eds) Elastic wave and ultrasonic nondestructive evaluation. North-Holland, Amsterdam
178. Fornberg B (1998) A practical guide to pseudospectral methods. Cambridge University Press, Cambridge
179. Fornberg B (1988) The pseudospectral method; accurate representation of interfaces in elastic wave calculations. Geophysics 53: 625-637
180. Krawczuk M, Ostachowicz W (2001) Spectral elements and genetic algorithm for crack detection in cantilever rod. Key Engineering Materials 204-205
181. Delsanto PP, Mignogna RB (1998) A spring model for the simulation of the propagation of ultrasonic pulse through imperfect contact interfaces. Journal of Acoustical Society of America: 104(5): 1-8
182. Yim H, Sohn Y (2000) Numerical simulation and visualization of elastic waves using mass-spring lattice model. IEEE Transactions on Ultrasonic, Ferroelectrics, and Frequency Control 47(3): 549-558
183. Delsanto PP, Whitcombe T, Chaskelis HH, Mignogna RB (1992) Connection machine simulation of ultrasonic wave propagation in materials. I: the one-dimensional case. Wave Motion 16: 65-80
184. Delsanto PP, Schechter RS, Chaskelis HH, Mignogna RB, Kline RB (1994) Connection machine simulation of ultrasonic wave propagation in materials. II: the two-dimensional case. Wave Motion 20: 295-314
185. Delsanto PP, Schechter RS, Mignogna RB (1997) Connection machine simulation of ultrasonic wave propagation in materials. III: the three-dimensional case. Wave Motion 26: 329-339
186. Lee BC, Staszewski WJ (2002) Local interaction modelling for acousto-ultrasonic wave propagation. In: Proceedings of the SPIE's 9th International Symposium on Smart Structures and Materials, Conference on Modeling, Signal Processing and Control in Smart Structures, San Diego, California. Paper No. 4693-32
187. Lee BC, Staszewski WJ (2003) Modelling of Lamb waves for damage detection in metallic structures. Part II: Wave interaction with damage. Smart Structures and Materials : 12(5): 815-824
188. Ihlenburg F (1998) Finite element analysis of acoustic scattering. Springer-Verlag, New York
189. Auld BA (1977) Symmetrical Lamb wave scattering at a symmetrical pair of thin solids. In: Ultrasonics Symposium Proceedings 247-253
190. Wang L, Shen J (1997) Scattering of elastic waves by a crack in a isotropic plate. Ultrasonics 35: 451-457

191. Norris AN, Vemula C (1995) Scattering of flexural waves in thin plates. Journal of Sound and Vibration 181: 115-125
192. Rokhlin SI (1981) Resonance phenomena of Lamb wave scattering by a finite crack in a solid plate. Journal of the Acoustical Society of America 69: 922-930
193. McKeon JCP, Hinders MK (1999) Lamb wave scattering from a through hole. Journal of Sound and Vibration 224(5): 843-862
194. Kim YW, Varadan VV, Varadan VK (1996) Modified null field method for elastic wave scattering from a partially debonded fiber - SH-waves. Composites Interfaces 3: 381-400
195. Kim YW, Varadan VV, Varadan VK (1997) Modified null field method for elastic wave scattering from a partially debonded fiber - P- and SV-waves. Composites Interfaces 4: 177-196
196. Shindo Y, Minamida K, Narita F (2002) Antiplane shear wave scattering from two curved interface cracks between a piezoelectric fiber and an elastic matrix. Smart Materials and Structures 11(4): 534-540
197. Verdict GS, Gien PH, Burger CP (1992) Finite element study of Lamb wave interactions with holesand through thickness defects in thin metal plates. In: Thompson DO, Chimenti DE (eds) Review of progress in quantitative nondestructive evaluation. Plenum Press, New York, 11: 97-104
198. (1992) Transient scattering of Rayleigh-Lamb waves by surface-breaking and buried cracks in a plate. In: Thompson DO, Chimenti DE (eds) Review of progress in quantitative nondestr uctive evaluation. Plenum Press, New York, 11: 73-80
199. Lin S, Kawashima K, Ito T (2000) Wave propagation analysis by finite element method for flaw sizing of circular pipes. In: Proceedings of the 15th World Congress on Nondestructive Testing, Rome, Italy
200. Lee BC, Staszewski WJ (2002) Modelling of acousto-ultrasonic wave interactions with defects in metallic structures. In: Proceedings of the International Conference on Noise and Vibration Engineering, ISMA 2002, Leuven, Belgium
201. Lee BC, Staszewski WJ (2003) Lamb wave interactions with structural defects - modelling and simulations. In: Proceedings of the SPIE's 10th International Symposium on Smart Structures and Materials, Conference on Modeling, Signal processing and Control, San Diego, California. Paper No. 5049-22.
202. Ostachowicz W, Krawczuk M (2001) On modelling of structural stiffness loss due to damage. In: Proceedings of the 4th International Conference on Damage Assessment of Structures (DAMAS 2001), Cardiff, UK 185-199
203. Staszewski W (2000) Advanced data pre-processing for damage identification based on pattern recognition. The International Journal of Systems Science 31(11): 1381-1396
204. Staszewski WJ, Worden K (2003) Signal processing for damage detection. In: Staszewski WJ, Boller C, Tomlinson GR (eds) Health Monitoring of Aerospace Structures. Wiley, Chichester
205. Rajagopalan C, Kalyanasundaram P, Raj B (1996) Uncertainty management in knowledge based systems for nondestructive testing - an example from ultrasonic testing. In: Proceedings of the 14th World Congress on NDT, New Delhi, India. 2127-2132
206. Hemez FM, Robertson AN, Cundy AL (2003) Uncertainty quantification and model validation for damage detection. In: Proceedings of the 4th International Workshop on Structural Health Monitoring, Stanford, California. 575-582

207. Farrar CF, Sohn H, Hemez FM, Anderson MC, Bement MT, Cornwell PJ, Doebling SW, Lieven N, Robertson AN, Schultze JF (2003) Damage prognosis: current status and future work. Technical report No. LA-14051-MS, Los Alamos Laboratory, Los Alamos, New Mexico
208. Farrar CF, Hemez F, Park G, Robertson AN, Sohn H, Williams TO (2003) A coupled approach to developing damage prognosis solutions. In: Proceedings of the International Conference on Damage Assessment of Structures (DAMAS-2003), Southampton, UK. 289-304
209. Farrar CF, Robertson AN (2003) An introduction to the Los Alamos damage prognosis project. In: Proceedings of the 4th International Workshop on Structural Health Monitoring, Stanford, California. 561-566
210. Staszewski WJ, Boller C, Tomlinson GR (eds) (2003) Health monitoring of aerospace structures: smart sensor technologies and signal processing. Wiley, Chichester
211. Web page: `www.ndt.net/article/az/ut/lamb.htm`
212. Hurlebaus S, Niethammer M, Jacobs LJ, Valle C (2001) Automated methodology to locate notches with Lamb waves. Acoustic Research Letters Online 2(4)
213. Fromme P, Sayir MB (2002) Measurement of the scattering of a Lamb wave by a through hole in a plate Journal of the Acoustical Society of America 111(3): 1165-1170
214. Schwarz WG, Read ME, Kremer MJ, Hinders MK, Smith BT (1999) Lamb wave tomographic imaging system for aircraft structural health assessment. In: Proceedings of the SPIE Conference on Nondestructive Evaluation of Aging Aircraft, Airports, and Aerospace Hardware III, SPIE 3586-292
215. Halabe UB, Franklin R (2001) Fatigue crack detection in metallic members using ultrasonic Rayleigh waves with time and frequency analyses. Material Evaluation 59(3): 424-431
216. Fromme P, Sayir MB (2002) Detection of cracks at rivet holes using guided waves. Ultrasonics 40(1-8): 199-203
217. Cordellos AD, Bell RO, Brummer SB (1969) Use of Rayleigh waves for the detection of stress-corrosion cracking (SCC) in aluminium alloys. Material Evaluation 27(4): 85-90
218. Gubbala R, Rao VS, Derriso MM (2003) Health monitoring of adhesively bonded repair of aircraft structures using wavelet transforms of Lamb wave signals. In: Proceedings of the 10th International Symposium on Smart Structures and Materials, Conference on Smart Structures and Integrated Systems, San Diego, California. Paper No. 5056-05
219. Ihn JB, Chang FK (2003) A smart patch for monitoring crack growth in metallic structures underneath bonded composite repair patches. In: Proceedings of the American Society for Composites 17th Technical Conference
220. Leutenegger TF (2002) Detection of defects in cylindrical structures using a time reverse numerical simulation method. PhD Thesis No. 14833. Swiss Federal Institute of Technology, Zurich
221. Salzburger HJ, Dobmann G, Mohrbacher H (2001) Quality control of laser welds of tailored blanks using guided waves and EMATs. IEE Proceedings, Science, Measurement and Technology 148(4): 143-148
222. Sun KJ (1991) Application of guided waves to delamination detection. In: Proceedings of the 18th Review of Progress in Qualitative Nondestructive Evaluation, Brunswick, Maine. 1213-1219

223. Kress KP (2003) Integrated imaging ultrasound SWISS. In: Proceedings of the 4th International Workshop on Structural Health Monitoring, Stanford, California. 852-859
224. Rose JL, Hay T (2000) Skin to honeycomb core delamination detection with guided waves. In: Proceedings of the 15th World Congress on NDT, Rome, Italy. Paper No. 271: `www.ndt.net/article/wcndt00/papers/idn271/idn217.htm`
225. Vine K (1999) The non-destructive testing of adhesive joints for environmental degradation. PhD Thesis. Department of Mechanical Engineering. Imperial College London
226. Rose JL Soley L (2000) Ultrasonic guided waves for the detection of anomalies in aircraft components. Materials Evaluation 50(9): 1080-1086
227. Dalton R (2000) Propagation of Lamb wave in metallic aircraft fuselage structure. PhD Thesis. Department of Mechanical Engineering. Imperial College London
228. Dalton RP, Cawley P, Lowe MJS (2001) The potential of guided waves for monitoring large areas of metallic aircraft fuselage structures. Journal of Non-Destructive Evaluation 20: 29-46
229. Rose JL, Soley LE, Hay T, Agarwala VS (2000) Ultrasonic guided waves for hidden corrosion detection in naval aircraft. In: Proceedings of the CORROSION NACExpo 2000 - 55th Annual Conference & Exposition, Orlando, Florida
230. Murfin AS, Dewhurst RJ (2002) Estimation of wall thinning in mild steel using laser ultrasound Lamb waves and a non-steady-state photo-emf detector. Ultrasonics 40(1-8): 777-781
231. Barshinger J, Rose JL, Avioli Jr. MJ (2002) Guided wave resonance tuning for pipe inspection. Journal of Pressure Vessel Technology 124: 303-310
232. Alleyne DN, Lank AM, PJ Mudge, Cawley P (1996) The Lamb wave inspection of chemical plant pipework. In: Proceedings of the 14th World Congress on NDT, New Delhi, India 4: 2303-2306
233. Park MH, Kim IS, Yoon YK (1996) Ultrasonic inspection of pipes using Lamb waves. NDT & E International 29(1): 13-20
234. Kwun H, Holt AE (1995) Feasibility of under-lagging corrosion detection in steel pipes using the magnetostrictive sensor technique. NDT & E International 28(4): 211-214
235. Chahbaz A, Mustafa V, Brassard M, Hay DR (1998) Inspection of compressed gas cylinder using ultrasonic guided waves. In: Proceedings of the 1st Pan American Conference for Nondestructive Testing - PACNDT'98i, Toronto, Canada, 4: 2
236. Beard MD, Lowe MJS Cawley P (2002) Development of a guided wave inspection technique for rock bolts. In: Thompson DO, Chimenti DE (eds) Review of Progress in Quantitative NDE, American Institute of Physics 21: 1318-1325
237. Beard MD, Lowe MJS Cawley P (2003) Non-destructive testing of rock bolts using guided ultrasonic waves. International Journal of Rock Mechanics and Mining Sciences 40: 527-536
238. Cawley P (2002) Practical long range guided wave inspection - applications to pipes and rail. In: Proceedings of the National Seminar of the Indian Society for Non-Destructive Testing (NDT-2002), Chennai, India. Paper No. 45
239. Wilcox P, Alleyne DN, Pavlakovic B, Evans MJ, Vine K, Cawley P, Lowe MJS (2003) Long range inspection of rail using guided waves. In: Thompson DO, Chimenti DE (eds) Review of Progress in Quantitative NDE, American Institute of Physics 22: 236-243

240. Cawley P, Lowe M, Wilcox P, Alleyne D, Pavlakovic B, Evans M, Vine K (2003) Long range inspection of rail using guided waves. Insight 45(6): 413-420
241. Armitage PR, Horwood JM (2003) The role of surface waves in assessing structural damage. In: Proceedings the 5th International Conference on Damage Assessment of Structures (DAMAS 2003), Southampton, UK 467-474
242. Rose JL, Avioli MJ (2000) Elastic wave analysis for broken rail detection. In:edings of the 15th World Congress on Nondestructive Testing, Rome. Paper No. 270: `www.ndt.net/article/wcndt00/papers/idn270/idn270.htm`
243. Pavlakovic B (1998) Leaky guided ultrasonic waves in NDT. PhD Thesis. Department of Mechanical Engineering. Imperial College London
244. Beard MD, Lowe MJS, Cawley P (2003) Inspection of steel tendons in concrete using guided waves. AIP Conf Proc 657 1139
245. Song WJ, Rose JL, Whitesel H (2002) Detection of damage in a ship hull using ultrasonic guided waves. Review of Progress in Quantitative NDE 21: 173-178
246. Song WJ, Rose JL, Whitesel H (2003) An ultrasonic guided wave technique for damage testing in a ship hull. Materials Evaluation 22(1): 94-98
247. Long RS, Vine K, Lowe MJS, Cawley P (2001) Monitoring of acoustic waves propagation in buried cast iron water pipes. In: Thompson DO, Chimenti DE (eds) Review of Progress in Quantitative NDE, American Institute of Physics 20: 1202-1209
248. Lowe MJS (1992) Plate waves for NDT of diffusion bonded titanium. PhD Thesis. Department of Mechanical Engineering. Imperial College London
249. Rose JL, Zhu W, Zaidi M (1998) Ultrasonic NDT of titanium diffusion bonding with guided waves. Material Evaluation 56(4): 535-539
250. Chan CW (1996) The ultrasonic non-destructive evaluation of welds in plastic pipes. PhD Thesis. Department of Mechanical Engineering. Imperial College London
251. Seale MD, Smith BT, Prosser WH, Master JE (1994) Lamb wave response of fatigued composite samples. Review of Progress in Quantitative Nondestructive Evaluation 13B: 1261-1266
252. Staszewski WJ, Pierce SG, Worden K, Philp WR, Tomlinson GR, Culshaw B (1997) Wavelet signal processing for enhanced Lamb wave defect detection in composite plates using optical fibre detection. Optical Engineering 36(7): 1877-1888
253. Worden K, Pierce SG, Manson G, Philp WR, Staszewski WJ, Culshaw B (2000) Detection of defects in composite plates using Lamb waves and novelty detection. International Journal of Systems Science 31(11): 1397-1409
254. Rose JL, Hay T, Agarwala VS (2000) Skin to honeycomb core delamination detection with guided waves. In: Proceedings of the 4th Joint DoD/FAA/NASA Conference on Aging Aircraft, St. Louis, MO
255. Yung YC, Kundu T, Ehsani M (2001) Internal discontinuity detection in concrete by Lamb waves. Material Evaluation 59(3): 418-423
256. Chiu WK, Galea SC, Koss LL, Rajic N (2000) Damage detection in bonded repairs using piezoceramics. Smart materials and Structures 9: 466-575
257. Luangvilai K, Punurai W, Jacobs LJ (2002) Guided Lamb wave propagation in composite plate/concrete component. Journal of Engineering Mechanics 128(12): 1337-1341
258. Park JP, Chang FK (2003) Built-in detection of impact damage in multilayered thick composite structures. In: Proceedings of the 4th International Workshop on Structural Health Monitoring, Stanford, California. 1391-1398

259. Na WB, Kundu T, Ehsani ME (2002) Ultrasonic guided waves for steel bar concrete interface testing. Material Evaluation 60(3): 437-444
260. Na WB, Kundu T, Ehsani MR (2003) Lamb waves for detecting delamination between steel bars and concrete. Computer-Aided Civil and Infrastructure Engineering 18: 58-63
261. Kawamoto S, Williams RS (2002) Acoustic Emission and acousto-ultrasonic techniques for wood and wood-based composites. Technical Report No. FPL-GTR-134. Forest Products Laboratory, United States Department of Agriculture
262. Grimberg R, Savin A, Lupu A, Rotundu C, Iancu L (2000) A Method to Determine the Debonding Zones in Multilayers Wood Materials. In: Proceedings of the 15th World Congress in NDT, Rome, Italy: `www.ndt.net/article/wcndt00/papers/ind5/ind5.htm`
263. V. Singh (2003) Active approach to damage detection in composite structures. MEng Thesis. Department of Mechanical Engineering, Sheffield University.

Controllability and Asymptotic Problems in Distributed Systems

Józef Joachim Telega

Institute of Fundamental Technological Research PAS
Świętokrzyska 21, 00–049 Warsaw, Poland
jtelega@ippt.gov.pl

1 Introduction

In our previous comprehensive papers [53, 54] the results pertaining to controllability of wave equations, linear and nonlinear vibrating solids and structures as well as heat equations were synthesized. However, asymptotic problems involving a small parameter ε intended to tend to zero were out of scope of these papers.

The aim of the present paper is to consider controllability problems where such a parameter appears. It is convenient to divide these problems into five classes:

1. singular perturbation problems,
2. "thin" domains (think of vibrating thin elastic plates),
3. homogenisation — including structures with periodically distributed patches,
4. perturbation of domains.

Among problems involving a small parameter one should also mention penalisation and numerical methods, cf. [34, 56, 57] and the relevant references therein. For instance, Zuazua [56] discussed recent developments pertaining to the numerical approximation of the linear wave equation. In this case, the interaction of waves with the numerical mesh produces spurious high frequency oscillations. It can also be proved that there are initial data for which the controls of the discrete models diverge as the mesh-size tends to zero.

Singular perturbation problems constitute a significant class od differential equations, not only in the context of controllability, see for instance Rodriguez-Bernal and Zuazua [39, 40] and the references therein.

Although von Kármán plates are also thin plates, yet treating them from the very beginning as two-dimensional structures we avoid the construction of the asymptotic model of such a plate. However, then the piezoelectric patches have to be assumed over the mid-plane of the plate. To distribute these patches over lower and upper faces of the plate one would have to consider a thin 3D domain with the patches distributed over the lower and upper faces and then pass asymptotically with thickness to zero, cf. [3, 11, 45].

The plan of the paper is as follows. Since the Hilbert Uniqueness Method (HUM), due to Lions [33, 34] plays a principal role in our considerations, therefore in Section 2 we recall its essential points. In Section 3 controllability problems for equations with singular perturbations are examined. Control of the wave and linear elasticity equations posed in thin domains are discussed in Section 4. Section 5 is concerned with a combination of homogenisation (highly heterogenous media) and controllability. Comments on von Karmán plates with periodically distributed piezoelectric patches and domains with micropores are provided in Section 6.

2 The Idea of HUM

Let Ω be a bounded domain in $\mathbb{R}^3$ (or $\mathbb{R}^n$, $n = 1, 2$ or 3) with a sufficiently smooth boundary $\Gamma = \partial\Omega$. We set $\dot{y} = \frac{\partial y}{\partial t}$, etc. Time interval is denoted by $(0, T)$, $T > 0$. Consider the following, simple wave equation:

$$\ddot{y} - \Delta y = 0 \quad \text{in } Q = \Omega \times (0, T) \tag{1}$$

$$y(0) = y^0, \quad \dot{y}(0) = y^1 \quad \text{in } \Omega \tag{2}$$

$$y = \begin{cases} v & \text{on } \Sigma_0 = \Gamma_0 \times (0, T) \\ 0 & \text{on } \Sigma \setminus \Sigma_0 \end{cases} \tag{3}$$

We shall often use the simple notation: $y(t) = \{y(x, t) | x \in \Omega\}$. By Γ_0 we denote a part of Γ. The case $\Gamma_0 = \Gamma$ is not precluded.

The problem we are interested in consists in driving system (1)–(3) to the desired final state, for instance

$$y(T; v) = \dot{y}(T; v) = 0 \quad \text{in } \Omega \tag{4}$$

Consider now the essential steps of the HUM.

(i) First, the homogeneous wave equation is examined:

$$\ddot{\Phi} - \Delta\Phi = 0 \quad \text{in } Q, \tag{5}$$

$$\Phi(0) = \Phi^0, \quad \dot{\Phi}(0) = \Phi^1 \quad \text{in } \Omega, \tag{6}$$

$$\Phi = 0 \quad \text{on } \Sigma. \tag{7}$$

(ii) The backward system is solved:

$$\begin{aligned} &\ddot{\Psi} - \Delta\Psi = 0 \quad \text{in } Q, \\ &\Psi(T) = \dot{\Psi}(T) = 0 \quad \text{in } \Omega, \\ &\Psi = \begin{cases} \frac{\partial\Phi}{\partial \mathbf{n}} & \text{on } \Sigma. \\ 0 & \text{on } \Sigma \setminus \Sigma_0. \end{cases} \end{aligned} \tag{8}$$

where $\mathbf{n}$ is the exterior unit normal to $\partial\Omega$.

(iii) A linear operator $\Lambda(\Phi^0, \Phi^1) = \{\dot{\Psi}(0), -\Psi(0)\}$ is defined. It can be shown that Λ is an isomorphism, and $\Lambda = \Lambda^*$, i.e. Λ is a symmetric operator. For details the reader is referred to Lions [33, 34].
Finally, we conclude that

$$\Lambda(\Phi^0, \Phi^1) = \{y^1, -y^0\} \tag{9}$$

possesses a unique solution $\{\Phi^0, \Phi^1\}$ for each pair of initial data $\{y^0, y^1\}$ and the control v appearing in (3) is given by

$$v = \frac{\partial \Phi}{\partial \mathbf{n}} \quad \text{on } \Sigma_0. \tag{10}$$

As usual in the mathematical analysis of controllability problems, also the HUM has to be cast in a proper setting of functional spaces. This point will become clear in our further developments.

3 Singular Perturbation Problems

Chapter 3 in the book by Lions [33] constitutes a good introduction to exact controllability and singular perturbation problems, cf. also Lions [32]. This author considers a model problem:

$$\ddot{y}_\varepsilon + \varepsilon \Delta^2 y_\varepsilon - \Delta y_\varepsilon = 0 \qquad \text{in } Q, \tag{11}$$

where $\varepsilon > 0$ is intended to tend to zero. Both Dirichlet and mixed controls on Σ are considered. For the Dirichlet control, duality has also been considered.

Fabre [20] studied the following exact controllability problem:

$$\begin{aligned} \ddot{y}_\varepsilon - \Delta y_\varepsilon &= v_\varepsilon \chi_{\omega_\varepsilon \times (0,T)} && \text{in } Q, \\ y_\varepsilon(0) &= 0 && \text{on } \Sigma, \\ y_\varepsilon(0) = y^0, \quad \dot{y}_\varepsilon(0) &= y^1 && \text{in } \Omega. \end{aligned} \tag{12}$$

We are interested in driving the system to rest as $\varepsilon \to 0$, and not only for each $\varepsilon > 0$. Here $\chi_{\omega_\varepsilon \times (0,T)}$ stands for the characteristic function of $\omega_\varepsilon \times (0, T)$ and ω_ε is defined as follows. Let Ω be a bounded domain in $\mathbb{R}^n$ with $\Gamma = \partial\Omega$ of class C^3. If Γ_0 is an open subset of Γ, we set $\Sigma_0 = \Gamma_0 \times (0, T)$. For $\varepsilon > 0$, ω_ε denotes the neighbourhood of Γ_0 defined by

$$\omega_\varepsilon = \left(\bigcup_{\mathbf{x} \in \Gamma_0} B(\mathbf{x}, \varepsilon) \right) \cap \Omega. \tag{13}$$

Here $B(\mathbf{x}, \varepsilon)$ is the ball with the centre $\mathbf{x}$ and the radius ε.

An interesting feature of passing with ε to zero combined with the HUM is that the limit system has the following form:

$$\begin{aligned}
&\ddot{y} - \Delta y = 0 && \text{in } Q, \\
&y = -\frac{1}{3}\frac{\partial \varphi}{\partial \mathbf{n}} \in L^2(\Sigma_0) && \text{on } \Sigma_0 \text{ and } y = 0 \quad \text{on } \Sigma \setminus \Sigma_0, \\
&y(0) = y^0, \quad \dot{y}(0) = y^1 && \text{in } \Omega, \\
&y(T) = 0, \quad \dot{y}(T) = 0 && \text{in } \Omega.
\end{aligned} \tag{14}$$

Here φ is obtained by passing to zero in the backward system. In her thesis Fabre [20] has also considered the Schrödinger equation and simplified plate equation.

Tcheugoué-Tébou [50] proved observability inequalities (essential in applying the HUM) for the wave equations with perturbations of order $4p$ and $4p+2$, $p \in \mathbb{N} \setminus \{0\}$; $\mathbb{N}$ stands for the set of natural numbers.

3.1 Singularly Perturbed Damped Wave Equation and Null Controllability of the Heat Equation

Let Ω be a bounded domain of class C^∞ of $\mathbb{R}^n$ (n — any space dimension; physically $n = 1$, 2 or 3). Consider the controllability of the following damped, singularly perturbed wave equation [37]:

$$\begin{aligned}
\varepsilon\ddot{u} - \Delta u + \dot{u} &= v\chi_\omega && \text{in } Q = \Omega \times (0,T), \\
u &= 0 && \text{on } \Sigma = \partial\Omega \times (0,T), \\
u(0) = u^0, \quad \dot{u}(0) &= u^1 && \text{in } \Omega.
\end{aligned} \tag{15}$$

Here χ_ω stands for the characteristic function of the open subset $\omega \subset \Omega$ where the control v is supported. This subset is defined as follows. Let $\mathbf{x}_0$ be *any given* point of $\mathbb{R}^n$. After Lions [33] we introduce the subset $\Gamma(\mathbf{x}^0)$ of $\Gamma = \partial\Omega$ defined by

$$\Gamma(\mathbf{x}^0) = \left\{ \mathbf{x} \in \Gamma \,\middle|\, (\mathbf{x} - \mathbf{x}^0) \cdot \mathbf{n}(\mathbf{x}) > 0 \right\}. \tag{16}$$

If $\mathcal{O}$ denotes a neighfourhood of $\Gamma(\mathbf{x}^0)$ in $\mathbb{R}^n$, we set $\omega = \Omega \cap \mathcal{O}$, cf. Fig. 1.

The formal limit when ε tends to zero of $(15)_1$ is the controlled heat equation:

$$\begin{aligned}
\dot{u} - \Delta u &= v\chi_\omega && \text{in } Q, \\
u &= 0 && \text{on } \Sigma, \\
u(x,0) &= u^0(x) && \text{in } \Omega.
\end{aligned} \tag{17}$$

Let $\varepsilon > 0$ be fixed. Denote by u_ε the solution of (15). We are interested in exact controllability of this system. In fact, since the system is linear, the problem is easily reduced to null controllability. More precisely, the problem consists in finding $v_\varepsilon \in L^2(\omega \times (0,T))$ such that the solution of (15) satisfies

$$u_\varepsilon(\mathbf{x},T) = \dot{u}(\mathbf{x},T) \equiv 0 \qquad \text{in } \Omega. \tag{18}$$

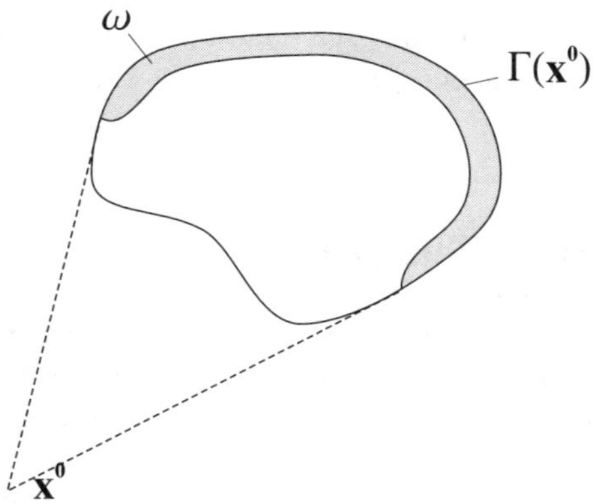

Fig. 1.

The second problem is to find the limit as $\varepsilon \longrightarrow 0$. The answer to both problems is positive and summarized as follows.

Theorem 1. *Assume that Ω and ω are as introduced above. Let $T > 0$. Then, there exists $\varepsilon(T) > 0$ such that for any $0 < \varepsilon < \varepsilon(T)$ system (15) is exactly controllable in time T ((18) is satisfied) and, moreover, there exists a constant $C(\varepsilon, T)$ bounded as $\varepsilon \longrightarrow 0$, such that*

$$\|v_\varepsilon\|_{L^2(\omega\times(0,T))} \leq C(\varepsilon, T)\|\{u^0, \sqrt{\varepsilon}u^1\}\|_{H_0^1(\Omega)\times L^2(\Omega)}$$

for each $\{u^0, u^1\} \in H_0^1(\Omega) \times L^2(\Omega)$.

Furthermore, for any $\{u^0, u^1\} \in H_0^1(\Omega) \times L^2(\Omega)$ fixed the controls v_ε of (15) can be chosen such that

$$v_\varepsilon \longrightarrow v \quad \text{in } L^2(\omega \times (0,T)) \text{ as } \varepsilon \longrightarrow 0, \tag{19}$$

v being a null control for the limit heat equation (17) with the initial temperature u^0.

Remark 1. The presence of damping in system (15) allows to consider T arbitrarily small, cf. Lions [34], López et al. [37], Telega [54], Telega and Bielski [53].

Obviously, for small T the cost of control is higher. In the case of exact controllability of the classical wave equation (without damping) the instant of time T is bounded from below by a positive quantity. We recall that in the classical case waves propagate with finite speed, hence also T has to be finite.

The proof of Theorem 1 is lengthy and involves many technical details. It is given by López et al. [37]. In essence, the main difficulty consists in finding the observability inequality for solutions of the backward system (the adjoint system). To this end, global Carleman's inequalities are derived, being extensions of those obtained for simpler cases by Fursikov and Imanuvilov [23].

Remark 2. HUM is based on the idea that observability of the adjoint (homogeneous) system is sufficient and necessary for the exact controllability, cf. Komornik [28, p. 56].

4 Controllability in Thin Domains

The aim of the presented section is to present three controllability problems in thin domains pertaining to: (i) internal and boundary exact controllability of the wave equation, (ii) exact controllability for a linear elastic solid, and (iii) null controllability of heat equations.

For results pertaining to thin cellular structures with periodically distributed holes the reader is referred to [48].

4.1 Wave Equations in Thin Domains

From the point of view of controllability and stabilization, linear and nonlinear wave equations seem to be tho most often studied equations, cf. Fursikov and Imanuvilov [23], Komornik [28], Lasiecka and Triggani [29], Lions [33, 34], Micu and Zuazua [38], Telega [54], Telega and Bielski [53] and references therein. For earlier results on exact controllability in thin domains the reader is referred to the book by Lions [34].

Distributed Internal Control

Let $\varepsilon > 0$ be a small parameter and $\Omega^\varepsilon = \Omega \times \left(-\frac{\varepsilon}{2}, \frac{\varepsilon}{2}\right)$, where Ω is a bounded, sufficiently regular domain of $\mathbb{R}^2$, cf. [21, 22].

By v we denote a control function distributed in Ω^ε such that for given initial conditions $\{y^0, y^1\}$ (in appropriate Hilbert spaces), if y is the solution to:

$$\begin{aligned} \ddot{y} - \Delta y = v \qquad & \text{in } \Omega^\varepsilon \times (0,T), \\ \frac{\partial y}{\partial \mathbf{n}} = 0 \qquad & \text{on } \Omega \times \left\{\frac{\pm\varepsilon}{2}\right\} \times (0,T), \\ y = 0 \qquad & \text{on } \left[\Gamma \times \left(-\frac{\varepsilon}{2}, +\frac{\varepsilon}{2}\right)\right] \times (0,T), \\ y(0) = y^0, \quad \dot{y}(0) = y^1 \qquad & \text{in } \Omega^\varepsilon, \end{aligned} \tag{20}$$

then

$$y(T) = \dot{y}(T) = 0 \qquad \text{in } \Omega^\varepsilon \quad \text{(null controllability).} \tag{21}$$

We recall that $\Gamma = \partial\Omega$ and

$$\dot{y} = \frac{\partial y}{\partial t}, \quad y(t) = \{y(\mathbf{x}, t) = y(x_1, x_2, x_3, t) | \mathbf{x} \in \Omega^\varepsilon\}.$$

Obviously, both the solution $y(\mathbf{x},t)$ and $v(\mathbf{x},t)$ depend on ε. The domain Ω^ε depends on ε. It is convenient to perform the following transformation, cf. Ciarlet [11],

$$(x_1,x_2,x_3) \in \Omega^\varepsilon \longrightarrow (z_1,z_2,z_3) \in \Omega^1 \tag{22}$$

where

$$z_\alpha = x_\alpha, \quad \alpha = 1,2; \quad z_3 = x_3/\varepsilon.$$

Similarly, for a function f we set

$$f(x_1,x_2,x_3) = f_\varepsilon(z_\alpha, z_3) \tag{23}$$

Exact controllability problem in the fixed domain Ω^1

Applying transformations (22)–(23) we transform problem (20), (21), cf. [21, 22]. Now the null controllability problem consists in finding a control v_ε distributed in Ω^1 such that if y_ε satisfies

$$\begin{aligned}
\ddot{y}_\varepsilon - \Delta_\varepsilon y_\varepsilon = v_\varepsilon \qquad & \text{in } \Omega^1 \times (0,T),\\
\frac{\partial y_\varepsilon}{\partial \mathbf{n}} = 0 \qquad & \text{on } \Omega \times \left\{\pm\frac{1}{2}\right\} \times (0,T),\\
y_\varepsilon = 0 \qquad & \text{on } \left[\Gamma \times \left(-\frac{1}{2},\frac{1}{2}\right)\right] \times (0,T),\\
y_\varepsilon(0) = y_\varepsilon^0, \quad \dot{y}_\varepsilon(0) = y_\varepsilon^1 \qquad & \text{in } \Omega^1,
\end{aligned} \tag{24}$$

then

$$y_\varepsilon(T) = \dot{y}_\varepsilon(T) = 0 \qquad \text{in } \Omega^1, \tag{25}$$

with

$$\Delta_\varepsilon = \frac{\partial^2}{\partial z_1^2} + \frac{\partial^2}{\partial z_2^2} + \varepsilon^{-2}\frac{\partial^2}{\partial z_3^2}. \tag{26}$$

Exact controllability in Ω^1

We introduce the space

$$V = \left\{u \in H^1(\Omega^1) | u = 0 \quad \text{on } \Gamma \times \left(-\frac{1}{2},\frac{1}{2}\right)\right\}$$

equipped with the following norm

$$\|u\|_\varepsilon^2 = \int_{-\frac{1}{2}}^{\frac{1}{2}} \int_\Omega \left\{ \left(\frac{\partial u}{\partial z_1}\right)^2 + \left(\frac{\partial u}{\partial z_2}\right)^2 + \varepsilon^{-2}\left(\frac{\partial u}{\partial z_3}\right)^2 \right\} dz_1 dz_2 dz_3.$$

We set $V_\varepsilon = (V, \|\cdot\|_\varepsilon)$ and denote by V_ε' the dual space of V_ε.

Now we are in a position to apply HUM to the exact controllability problem (24), (25) posed on Ω^1. First we introduce φ_ε, a solution to the following homogeneous system:

$$
\begin{aligned}
\ddot{\varphi}_\varepsilon - \Delta_\varepsilon \varphi_\varepsilon &= 0 && \text{in } \Omega^1 \times (0,T),\\
\frac{\partial \varphi_\varepsilon}{\partial \mathbf{n}} &= 0 && \text{on } \Omega \times \left\{ \pm\frac{1}{2} \right\} \times (0,T),\\
\varphi_\varepsilon &= 0 && \text{on } \left[\Gamma \times \left(-\frac{1}{2}, \frac{1}{2} \right) \right] \times (0,T),\\
\varphi_\varepsilon(0) = \varphi_\varepsilon^0, \quad \dot{\varphi}_\varepsilon(0) &= \varphi_\varepsilon^1 && \text{in } \Omega^1.
\end{aligned}
\tag{27}
$$

Remark 3. If $\varphi_\varepsilon^0 \in L^2(\Omega^1)$ and $\varphi_\varepsilon^1 \in V_\varepsilon'$, then φ_ε solving (27) verifies

$$
\varphi_\varepsilon \in C\left([0,T]; L^2(\Omega^1)\right) \cap C^1([0,T]; V_\varepsilon').
$$

For definitions and properties of functional spaces appropriate for evolution problems the reader is referred to [11, 27, 28, 31, 33, 34] and relevant books cited therein.

The following result plays an important role.

Lemma 1. *There exist positive constants c_1 and c_2 independent of ε, such that*

$$
c_1 \left\{ \|\varphi_\varepsilon^0\|^2_{L^2(\Omega^1)} + \|\varphi_\varepsilon^1\|^2_{V_\varepsilon'} \right\}^{1/2} \leq \|\varphi\|_{L^2(\Omega^1 \times (0,T))}
\leq c_2 \left\{ \|\varphi_\varepsilon^0\|^2_{L^2(\Omega^1)} + \|\varphi_\varepsilon^1\|^2_{V_\varepsilon'} \right\}^{1/2}. \tag{28}
$$

□

We set

$$
F_\varepsilon = L^2(\Omega^1) \times V_\varepsilon', \quad F_\varepsilon' = V_\varepsilon \times L^2(\Omega^1) \tag{29}
$$

and define

$$
\left\| \{\varphi_\varepsilon^0, \varphi_\varepsilon^1\} \right\|_{F_\varepsilon} = \left\{ \int_0^T \int_{\Omega^1} (\varphi_\varepsilon)^2 d\mathbf{z} dt \right\}^{1/2}. \tag{30}
$$

From Lemma 1 we conclude that $\|\cdot\|_{F_\varepsilon}$ is a norm on F_ε *uniformly* equivalent to the norm of F_ε.

Consider now the adjoint or backward system:

$$
\begin{aligned}
\ddot{\Psi}_\varepsilon - \Delta_\varepsilon \Psi_\varepsilon &= -\varphi_\varepsilon && \text{in } \Omega^1 \times (0,T),\\
\frac{\partial \Psi_\varepsilon}{\partial \mathbf{n}} &= 0 && \text{on } \left[\Omega \times \left\{ \pm\frac{1}{2} \right\} \right] \times (0,T),\\
\Psi_\varepsilon &= 0 && \text{on } \left[\Gamma \times (-\frac{1}{2}, \frac{1}{2}) \right] \times (0,T),\\
\Psi_\varepsilon(T) = 0, \quad \dot{\Psi}_\varepsilon(T) &= 0 && \text{in } \Omega^1,
\end{aligned}
\tag{31}
$$

where φ_ε in $(31)_1$ is a solution to (27). System (31) possesses a unique solution Ψ_ε satisfying:

$$
\Psi_\varepsilon \in C\left([0,T]; V_\varepsilon\right) \cap C^1([0,T]; L^2(\Omega^1)).
$$

Now we are in a position to define the linear and self-adjoint operator

$$\Lambda_\varepsilon : F_\varepsilon \longrightarrow F'_\varepsilon$$

by

$$\Lambda_\varepsilon\{\varphi^0_\varepsilon, \varphi^1_\varepsilon\} = \{\dot{\Psi}_\varepsilon(0), -\Psi_\varepsilon(0)\}.$$

Λ_ε is an *isomorphism* of F_ε on F'_ε. Hence the equation

$$\Lambda_\varepsilon\{\varphi^0_\varepsilon, \varphi^1_\varepsilon\} = \{y^1_\varepsilon, -y^0_\varepsilon\} \tag{32}$$

is uniquely solvable and the control function is given by

$$v_\varepsilon = -\varphi_\varepsilon. \tag{33}$$

Moreover, $y_\varepsilon = \Psi_\varepsilon$. Since $\Psi_\varepsilon(T) = 0$ and $\dot{\Psi}(T) = 0$ in Ω^1, therefore $y_\varepsilon(T) = \dot{y}_\varepsilon(T) = 0$ in Ω^1.

Remark 4. The control v_ε minimizes the functional

$$\int_0^T\int_\Omega u^2 d\mathbf{z}dt$$

on the set of admissible controls, i.e. satisfying (24).

Let us now pass to the limit problem as $\varepsilon \longrightarrow 0$. To this end we make the following assumption on the initial data y^0_ε and y^1_ε:

$$\begin{aligned} \|y^0_\varepsilon\|_\varepsilon \le c \quad \text{and} \quad y^0_\varepsilon &\rightharpoonup \tilde{y}^0 \text{ weakly in } H^1(\Omega), \\ y^1_\varepsilon &\rightharpoonup \tilde{y}^1 \text{ weakly in } L^2(\Omega^1). \end{aligned} \tag{34}$$

We recall that a sequence $\{f_\varepsilon\} \subset L^2(\Omega^1)$ converges weakly to $f \in L^2(\Omega)$, i.e.,

$$f_\varepsilon \rightharpoonup f \text{ weakly in } L^2(\Omega^1) \text{ as } \varepsilon \longrightarrow 0$$

if

$$\int_{\Omega^1} f_\varepsilon g d\mathbf{z} \longrightarrow \int_{\Omega^1} f g d\mathbf{z} \quad \forall g \in L^2(\Omega^1).$$

The main result is formulated as follows.

Theorem 2. *The control v_ε of the system* (24) *(obtained by HUM) is such that $v_\varepsilon \rightharpoonup \tilde{v}$ weakly in $L^2(\Omega^1 \times (0,T))$ as $\varepsilon \longrightarrow 0$ where $\tilde{v}$ is independent of z_3 and $\tilde{v}$ is the control function (exact) of:*

$$\begin{aligned} \ddot{\tilde{y}} - \Delta_2\tilde{y} &= \tilde{v} && \text{in } \Omega \times (0,T) \quad \text{(2D equation)}, \\ \tilde{y} &= 0 && \text{on } \Gamma \times (0,T), \\ \tilde{y}(0) = \tilde{y}^0, \quad \dot{\tilde{y}} &= \mathcal{M}(\tilde{y}^1) && \text{in } \Omega, \end{aligned} \tag{35}$$

where

$$\mathcal{M}(\tilde{y}^1)(z_\alpha) = \int\limits_{-1/2}^{1/2} \tilde{y}^1(z_\alpha, z_3)dz_3, \quad \Delta_2 = \frac{\partial^2}{\partial z_1^2} + \frac{\partial^2}{\partial z_2^2}.$$

Moreover, we have

$$y_\varepsilon \rightharpoonup \tilde{y} \text{ weak-* in } L^\infty(0,T;H^1(\Omega^1)), \quad \dot{y}_\varepsilon \rightharpoonup \dot{\tilde{y}} \text{ weak-* in } L^\infty(0,T;L^2(\Omega^1)).$$

Comments

(i) It seems possible to extend Th. 2 to the wave equation in inhomogeneous and anisotropic media where,

$$\ddot{y} - \frac{\partial}{\partial x_i}\left(a_{ij}(\mathbf{x})\frac{\partial y}{\partial x_j}\right) = v \qquad \text{in } \Omega^\varepsilon \times \left(-\frac{\varepsilon}{2}, \frac{\varepsilon}{2}\right). \tag{36}$$

(ii) Yan [55] considered the exact boundary controllability in the following case:

$$\ddot{y} - \Delta y = 0 \qquad \text{in } \Omega^\varepsilon \times (0,T);$$

$$y = \begin{cases} v_\varepsilon \text{ (control)} & \text{on } \Sigma_\varepsilon(\mathbf{x}^0) = \Gamma_\varepsilon(\mathbf{x}^0) \times (0,T), \\ 0 & \text{on } \Sigma_\varepsilon \setminus \Sigma_\varepsilon(\mathbf{x}^0); \end{cases}$$

$$\frac{\partial y}{\partial \mathbf{n}} = \begin{cases} w_\varepsilon \text{ (control)} & \text{on } \left[\Gamma \times \left\{+\frac{1}{2}\right\}\right] \times (0,T), \\ 0 & \text{on } \left[\Gamma \times \left(-\frac{1}{2}\right)\right] \times (0,T); \end{cases}$$

$$y(0) = y^0, \quad \dot{y}(0) = y^1 \qquad \text{in } \Omega^\varepsilon = \Omega \times (-\varepsilon/2, +\varepsilon/2).$$

Here the following notation is introduced: $\Gamma_\varepsilon(\mathbf{x}^0) = \Gamma(\mathbf{x}^0) \times (-\varepsilon/2, \varepsilon/2)$, $\mathbf{x}^0 \in \mathbb{R}^2$, $\Sigma_\varepsilon = \Gamma \times (-\varepsilon/2, \varepsilon/2) \times (0,T)$, $\mathbf{x}^0 \in \mathbb{R}^2$, and

$$\Gamma(\mathbf{x}^0) = \{\mathbf{x} \in \Gamma | \mathbf{m}(\mathbf{x}) \cdot \mathbf{n}(\mathbf{x}) > 0\}, \quad \mathbf{m}(\mathbf{x}) = \mathbf{x} - \mathbf{x}^0$$

Problem: Find controls v, w such that

$$y(T) = \dot{y}(T) = 0 \qquad \text{in } \Omega^\varepsilon \times (0,T).$$

Once again, the Hilbert Uniqueness Method was used. However, the exact boundary controllability problem is more complex than the problem of internal control. For instance, one cannot expect the following behaviour of the "classical" operator Λ_ε:

$$\Lambda_\varepsilon \longrightarrow \Lambda, \quad \varepsilon \longrightarrow 0, \text{ for a certain topology,}$$

where

$$\Lambda : F = H_0^1(\Omega) \times L^2(\Omega) \longrightarrow F' = H^{-1}(\Omega) \times L^2(\Omega),$$

is the isomorphism defined for the exact controllability in the Ω. To cope with the problem one, Yan [55] introduced a properly defined, modified operator Θ_ε, being an isomorphism. Unfortunately, the operator Θ_ε constructed by Yan is not symmetric and his control functions v_ε and w_ε are not minimisers of a cost functional, thus they are not optimal.

4.2 Exact Controllability of Thin Linear Elastic Solids and Asymptotic Limit for Plates

Let $\bar{\Omega}^\varepsilon = \bar{\omega} \times [-\varepsilon, \varepsilon]$, where ω is a polygonal domain in $\mathbb{R}^2$. In its undeformed state a thin plate occupies the closure of the domain Ω^ε, i.e. $\bar{\Omega}^\varepsilon$. By $\Gamma_\pm^\varepsilon = \partial\omega \times \{-\varepsilon, \varepsilon\}$ we denote the upper and bottom faces of the plate and $\Gamma_0^\varepsilon = \partial\omega \times (-\varepsilon, \varepsilon)$ is its lateral surface.

Consider the following 3D linear elasticity problem in the 'thin' domain Ω^ε, cf. [21, 22],

$$\begin{aligned} \rho^\varepsilon \ddot{u}_i^\varepsilon - \partial_j^\varepsilon \sigma_{ij}^\varepsilon(\mathbf{u}^\varepsilon) &= 0 && \text{in } Q^\varepsilon = \Omega^\varepsilon \times (0,T), \\ \mathbf{u}^\varepsilon &= \mathbf{\Psi}^\varepsilon && \text{on } \Sigma_0^\varepsilon = \Gamma_0^\varepsilon \times (0,T), \\ \boldsymbol{\sigma}^\varepsilon(\mathbf{u}^\varepsilon)\mathbf{n}^\varepsilon &= \mathbf{v}^\varepsilon && \text{on } \Sigma_\pm^\varepsilon = \Gamma_\pm^\varepsilon \times (0,T), \\ \mathbf{u}^\varepsilon(0) = \mathbf{u}_0^\varepsilon, \quad \dot{\mathbf{u}}^\varepsilon(0) &= \mathbf{u}_1^\varepsilon && \text{in } \Omega^\varepsilon. \end{aligned} \tag{37}$$

The vector $\mathbf{n}^\varepsilon = (n_i^\varepsilon)$ is the unit outward normal vector to $\partial\Omega^\varepsilon$, ρ^ε denotes the mass density and σ_{ij}^ε, $i,j = 1,2,3$, are the components of the stress tensor. The plate is isotropic and homogeneous:

$$\sigma_{ij}(\mathbf{u}^\varepsilon) = \lambda^\varepsilon e_{kk}^\varepsilon(\mathbf{u}^\varepsilon)\delta_{ij} + 2\mu^\varepsilon e_{ij}^\varepsilon(\mathbf{u}^\varepsilon) \tag{38}$$

where $e_{ij}^\varepsilon(\mathbf{u}^\varepsilon) = (\partial_j^\varepsilon u_i^\varepsilon + \partial_i^\varepsilon u_j^\varepsilon)/2$, $\partial_j^\varepsilon = \partial/\partial x_j^\varepsilon$, $\dot{f} = \partial f/\partial t$, and the Lamé constants λ^ε and μ^ε do not depend on $\mathbf{x}^\varepsilon$. We assume that

$$\lambda^\varepsilon = \lambda, \quad \mu^\varepsilon = \mu, \quad \rho^\varepsilon = \varepsilon^2 \rho \tag{39}$$

where ρ is obviously also independent of $\mathbf{x}^\varepsilon$.

The exact controllability problem for (37) is here formulated as follows: given the initial data $\{\mathbf{u}_0^\varepsilon, \mathbf{u}_1^\varepsilon\}$ in an energy space to be precised below, does there exist a time $T > 0$ and *control functions* $\mathbf{\Psi}^\varepsilon$ on Γ_0^ε and $\mathbf{v}^\varepsilon$ on $\Gamma_\pm^\varepsilon$, such that the unique solution $\mathbf{u}^\varepsilon$ to (37) achieves the equilibrium at the time T, i.e. $\mathbf{u}^\varepsilon(T) = \mathbf{0}$, $\dot{\mathbf{u}}^\varepsilon(T) = \mathbf{0}$?

To solve this problem, Figueiredo and Zuazua [21, 22] used the multiplier and HUM methods, cf. [28, 33, 34]. The next problem is to construct controls $\mathbf{\Psi}^\varepsilon$ and $\mathbf{v}^\varepsilon$, such that, when $\varepsilon \to 0$, the controlled solution $\mathbf{u}^\varepsilon$ (satisfying $\mathbf{u}^\varepsilon(T) = 0$ and $\dot{\mathbf{u}}^\varepsilon(T) = 0$) converges in a suitable topology to the controlled solution of the limit 2D plate. To achieve this goal, the asymptotic analysis as presented in Ciarlet [11] combined with Lions' approach [33, 34] is used [21, 22].

To this end we first perform suitable scaling in order to deal with a plate of thickness 2, thus independent of ε. We set $\Omega = \omega \times (-1,1)$, $\gamma = \partial w$, $\Gamma_\pm = \partial\omega \times \{-1,1\}$, $\Gamma_0 = \partial\omega \times (-1,1)$.

According to the asymptotic method [11], we introduce the scaled displacement $\mathbf{u}(\varepsilon) = (u_i(\varepsilon)) : \Omega \times (0,T) \to \mathbb{R}^3$ and the scaled initial data $\mathbf{u}_0(\varepsilon) = (u_{0i}(\varepsilon))$ and $\mathbf{u}_1(\varepsilon) = (u_{1i}(\varepsilon)) : \bar{\Omega} \to \mathbb{R}^3$ in the following way:

$$\begin{aligned} u^\varepsilon_\alpha(\mathbf{x}^\varepsilon, t) &= \varepsilon^2 u_\alpha(\varepsilon)(\mathbf{x}, t), \quad u^\varepsilon_3(\mathbf{x}^\varepsilon, t) = \varepsilon u_3(\varepsilon)(\mathbf{x}, t), \\ u^\varepsilon_{0\alpha}(\mathbf{x}^\varepsilon) &= \varepsilon^2 u_{0\alpha}(\varepsilon)(\mathbf{x}), \quad u^\varepsilon_{03}(\mathbf{x}^\varepsilon) = \varepsilon u_{03}(\varepsilon)(\mathbf{x}), \\ u^\varepsilon_{1\alpha}(\mathbf{x}^\varepsilon) &= \varepsilon^2 u_{1\alpha}(\varepsilon)(\mathbf{x}), \quad u^\varepsilon_{13}(\mathbf{x}^\varepsilon) = \varepsilon u_{13}(\varepsilon)(\mathbf{x}). \end{aligned} \tag{40}$$

Here $\mathbf{x}^\varepsilon = \pi^\varepsilon(\mathbf{x}) \in \bar{\Omega}^\varepsilon$, and $\pi^\varepsilon(\mathbf{x}) = \pi^\varepsilon(x_1, x_2, x_3) = (x_\alpha, \varepsilon x_3)$, $(x_i) \in \bar{\Omega}$.

Next, we define functional spaces:

$$\begin{aligned} H^1_{\Gamma_0} &= \left\{u \in H^1(\Omega) | u = 0 \text{ on } \Gamma_0\right\}, \quad V(\Omega) = [H^1_{\Gamma_0}(\Omega)]^3, \\ X &= \left\{g \in L^1(0,T; L^2(\Omega)) | \dot{g} \in L^1(0,T; [V(\Omega)]')\right\}. \end{aligned} \tag{41}$$

The spaces $H^1_{\Gamma_0}(\Omega)$ and $V(\Omega)$ are equipped with the usual norms whilst on X the norm is defined by

$$\|g\|^2_X = \|g\|^2_{L^1(0,T;L^2(\Omega))} + \|\dot{g}\|^2_{L^1(0,T;[V(\Omega)]')}.$$

The space $(X; \|\cdot\|_X)$ is a Banach space.

Taking into account (40) in (37), we readily derive the scaled linear elasticity problem:

$$\begin{aligned} \varepsilon^2 \rho \ddot{u}_\alpha - \partial_j \sigma_{\alpha j}(\varepsilon)(\mathbf{u}(\varepsilon)) &= 0 && \text{in } Q = \Omega \times (0,T), \\ \rho \ddot{u}_3(\varepsilon) - \partial_j \sigma_{3j}(\varepsilon)(\mathbf{u}(\varepsilon)) &= 0 && \text{in } Q, \\ \mathbf{u}(\varepsilon) &= \boldsymbol{\Psi}(\varepsilon) && \text{on } \Sigma_0 = \Gamma_0 \times (0,T), \\ \boldsymbol{\sigma}(\varepsilon)(\mathbf{u}(\varepsilon))\mathbf{n} &= \mathbf{v}(\varepsilon) && \text{on } \Sigma_\pm = \Gamma_\pm \times (0,T), \\ \mathbf{u}(\varepsilon)(0) = \mathbf{u}_0(\varepsilon), \quad \dot{\mathbf{u}}(\varepsilon)(0) &= \mathbf{u}_1(\varepsilon) && \text{in } \Omega. \end{aligned} \tag{42}$$

where $\partial_j = \partial/\partial x_j$. The scaled components of the stress tensor $\sigma_{ij}(\varepsilon)(\cdot)$ are specified by:

$$\begin{aligned} \sigma_{\alpha\beta}(\varepsilon)(\mathbf{u}(\varepsilon)) &= \lambda \left[e_{\gamma\gamma}(\mathbf{u}(\varepsilon)) + \frac{1}{\varepsilon^2} e_{33}(\mathbf{u}(\varepsilon))\right] \delta_{\alpha\beta} + 2\mu_{\alpha\beta}(\mathbf{u}(\varepsilon)), \\ \sigma_{\alpha 3}(\varepsilon)(\mathbf{u}(\varepsilon)) &= 2\mu \frac{1}{\varepsilon^2} e_{\alpha 3}(\mathbf{u}(\varepsilon)), \\ \sigma_{33}(\varepsilon)(\mathbf{u}(\varepsilon)) &= \frac{1}{\varepsilon^2} \left\{\lambda \left[e_{\alpha\alpha}(\mathbf{u}(\varepsilon)) + \frac{1}{\varepsilon^2} e_{33}(\mathbf{u}(\varepsilon))\right] + 2\mu \frac{1}{\varepsilon^2} e_{33}(\mathbf{u}(\varepsilon))\right\}. \end{aligned} \tag{43}$$

The vector $\mathbf{n} = (n_i)$ in $(42)_4$ is the outward unit normal to $\partial\Omega$.

According to HUM [33] we introduce the function

$$\mathbf{m}(\mathbf{x}) = \mathbf{x} - \mathbf{x}_0, \quad \forall \mathbf{x} \in \bar{\Omega}, \tag{44}$$

where $\mathbf{x}_0 = (x_{0\alpha}, 0)$ is a fixed point in ω, and the constants $R(\mathbf{x}_0)$ and $T(\varepsilon)$ by

$$R(\mathbf{x}_0) = \|\mathbf{x} - \mathbf{x}_0\|_{L^\infty(\Omega)}, \quad T(\varepsilon) = \varepsilon \frac{2\sqrt{\rho}}{\sqrt{\mu}} \max\left\{ R(\mathbf{x}_0), \frac{C(\Omega)}{R(\mathbf{x}_0)} \right\}. \tag{45}$$

Here $C(\Omega)$ is the constant of the trace map tr:$V \to L^2(\partial\Omega)^3$. By $\tau_j (j = 1, 2, 3)$ we denote the components of the tangential gradient on $\partial\Omega$.

The exact controllability result for the 3D plate identified with $\bar{\Omega}$ is formulated as follows [21, 22].

Theorem 3. *Let $0 < \varepsilon \le 1$ be held fixed. If $T > T(\varepsilon)$, then the elasticity system (42) is exactly controllable. More precisely, if $\mathbf{u}_0 \in L^2(\Omega)^3$, $\mathbf{u}_1 \in [V(\Omega)]'$, then there exist controls given by*

$$\begin{aligned} \Psi_i(\varepsilon) &= m_j n_j \frac{\partial \phi_i(\varepsilon)}{\partial \mathbf{n}} && \text{on } \Gamma_0 \times (0, T), \\ v_\alpha(\varepsilon) &= (m_j n_j)[\varepsilon^2 \rho \ddot{\phi}_\alpha(\varepsilon) - \tau_j \sigma_{\alpha j}(\varepsilon)(\boldsymbol{\phi}(\varepsilon))] && \text{on } \Gamma_\pm \times (0, T), \\ v_3(\varepsilon) &= (m_j n_j)[\rho \ddot{\phi}_3(\varepsilon) - \tau_j \sigma_{3j}(\varepsilon)(\boldsymbol{\phi}(\varepsilon))] && \text{on } \Gamma_\pm \times (0, T), \end{aligned} \tag{46}$$

where $\boldsymbol{\phi}(\varepsilon) = (\phi_i(\varepsilon))$ is the solution to the following homogeneous *3D elasticity problem:*

$$\begin{aligned} \varepsilon^2 \rho \ddot{\phi}_\alpha - \partial_j \sigma_{\alpha j}(\varepsilon)(\boldsymbol{\phi}(\varepsilon)) &= 0 && \text{in } Q = \Omega \times (0, T), \\ \rho \ddot{\phi}_3(\varepsilon) - \partial_j \sigma_{3j}(\varepsilon)(\boldsymbol{\phi}(\varepsilon)) &= 0 && \text{in } Q, \\ \boldsymbol{\phi}(\varepsilon) &= \mathbf{0} && \text{on } \Sigma_0 = \Gamma_0 \times (0, T), \\ \boldsymbol{\sigma}(\varepsilon)(\boldsymbol{\phi}(\varepsilon))\mathbf{n} &= \mathbf{v}(\varepsilon) && \text{on } \Sigma_\pm = \Gamma_\pm \times (0, T), \\ \boldsymbol{\phi}(\varepsilon)(0) = \boldsymbol{\phi}_0(\varepsilon), \quad \dot{\boldsymbol{\phi}}(\varepsilon)(0) &= \boldsymbol{\phi}_1(\varepsilon) && \text{in } \Omega, \end{aligned} \tag{47}$$

such that the solution $\mathbf{u}(\varepsilon)$ of (42) satisfies

$$\mathbf{u}(\varepsilon)(T) = \mathbf{0}, \quad \dot{\mathbf{u}}(\varepsilon)(T) = \mathbf{0}. \tag{48}$$

The initial control data $\{\boldsymbol{\phi}_0(\varepsilon), \boldsymbol{\phi}_1(\varepsilon)\}$ belong to $V(\Omega) \times L^2(\Omega)^3$ and verify:

$$\Lambda(\varepsilon)(\{\boldsymbol{\phi}_0(\varepsilon), \boldsymbol{\phi}_1(\varepsilon)\}) = \{\mathbf{u}_1(\varepsilon), -\mathbf{u}_0(\varepsilon)\}, \tag{49}$$

where $\Lambda(\varepsilon) : V(\Omega) \times L^2(\Omega)^3 \to [V(\Omega)]' \times L^2(\Omega)^3$ is the isomorphism given by HUM.

The proof of the above theorem is given in [22].

Remark 5.

(i) The optimal controllability time is

$$T^\star(\varepsilon) = 2\varepsilon\sqrt{\rho/\mu}R(\mathbf{x}_0)$$

and not the time $T(\varepsilon)$ specified by $(45)_2$. However, we have to consider time $T(\varepsilon)$ and not $T^\star(\varepsilon)$ in order to pass with ε to zero.

(ii) The controls $\boldsymbol{\Psi}(\varepsilon)$ and $\mathbf{v}(\varepsilon)$ acting on Γ_0 and $\Gamma_\pm$ have different structure. It can be shown that

$$u_i(\varepsilon) = m_k n_k \frac{\partial \phi_i(\varepsilon)}{\partial \mathbf{n}} \in L^2(\Gamma_0 \times (0,T))$$

since

$$\frac{\partial \phi_i(\varepsilon)}{\partial \mathbf{n}} \in L^2(0,T;L^2(\Gamma_0))$$

for every solution of (47) with initial data in $V(\Omega) \times L^2(\Omega)^3$.

(iii) The control $\mathbf{v}(\varepsilon)$ is to be understood in the sense of the following duality formula:

$$\langle \mathbf{v}(\varepsilon), \boldsymbol{\theta}\rangle = \int_{\Sigma_\pm} m_k n_k [\sigma_{ij}(\varepsilon)(\boldsymbol{\phi}(\varepsilon))e_{ij}(\boldsymbol{\theta}) - \varepsilon^2\rho\dot{\phi}_\alpha(\varepsilon)\dot{\theta}_\alpha - \rho\dot{\phi}_3(\varepsilon)\dot{\theta}_3]d\Gamma_\pm dt$$

for all $\boldsymbol{\theta} = (\theta_i)$ sufficiently smooth.

Limit behaviour as $\varepsilon \to 0$

The passage from 3D to plate problem is summarized as follows.

Theorem 4. *Assume that $T > 0$. Let $\{\mathbf{u}(\varepsilon)\}_{\varepsilon>0}$ be the sequence of controlled solutions of problem* (42) *(for $\varepsilon > 0$ small enough, such that $T(\varepsilon) < T$) with controls* (46). *Suppose that the corresponding sequence of initial data $\{\mathbf{u}_0(\varepsilon), \mathbf{u}_1(\varepsilon)\}$ verifies*

$$\exists C > 0, \ \ \|\{\mathbf{u}_1(\varepsilon), -\mathbf{u}_0(\varepsilon)\}\|_{[V(\Omega)]'\times L^2(\Omega)^3} \le C,$$

where C is independent of ε. Then there exists a subsequence $\{\mathbf{u}(\varepsilon_k)\}_{\varepsilon_k>0}$ of $\{\mathbf{u}(\varepsilon)\}_{\varepsilon>0}$ and functions $\{u_\alpha, u_3\}$ in the space $(X')^2 \times L^\infty(0,T;L^2(\Omega))$ such that, for any $\{f_\alpha, f_3\} \in X^2 \times L^1(0,T;L^2(\Omega))$

$$\langle u_i(\varepsilon_k), f_i\rangle \to \langle u_i, f_i\rangle \ \textit{as}\ \varepsilon_k \to 0.$$

The limit function $\mathbf{u} = (u_i)$ satisfies:

(i) $\mathbf{u} = (u_i)$ is a Kirchhoff-Love displacement, that is u_3 is independent of x_3 and $u_\alpha = \hat{u}_\alpha - x_3\partial_\alpha u_3$, where $\hat{u}_\alpha$ is independent of x_3.

(ii) u_3 *is the solution (in the transposition sense, cf. Lions [33]) of the following 2D plate problem*

$$
\begin{aligned}
2\rho\ddot{u}_3 + \frac{8\mu(\lambda+\mu)}{3(\lambda+2\mu)}\Delta^2 u_3 = 2\rho\ddot{\phi}_3 + \frac{8\mu(\lambda+\mu)}{\lambda+2\mu}\Delta^2\phi_3 &\quad \text{in } \omega\times(0,T),\\
u_3 = 0 &\quad \text{on } \partial\omega\times(0,T),\\
\frac{\partial u_3}{\partial \mathbf{n}} = m_\alpha n_\alpha \Delta\phi_3 &\quad \text{on } \partial\omega\times(0,T),\\
u_3(0) = \frac{1}{2}\int_{-1}^{+1} u_{03}dx_3, \quad \dot{u}_3(0) = \frac{1}{2}\int_{-1}^{+1} u_{13}dx_3 &\quad \text{in } \omega,\\
u_3(T) = 0, \quad \dot{u}_3(T) = 0 &\quad \text{in } \omega.
\end{aligned}
\tag{50}
$$

The couple $\{u_{03}, u_{13}\}$ *is the weak limit in the space* $L^2(\Omega)\times[H^1_{\Gamma_0}(\Omega)]'$ *of the sequence* $\{u_{03}(\varepsilon_k), u_{13}(\varepsilon_k)\}_{\varepsilon_k>0}$ *and the function* ϕ_3*, defining the controls in* (50)*, is the unique solution to*

$$
\begin{aligned}
2\rho\ddot{\phi}_3 + \frac{8\mu(\lambda+\mu)}{3(\lambda+2\mu)}\Delta^2\phi_3 = 0 &\quad \text{in } \omega\times(0,T),\\
\phi_3 = \frac{\partial\phi_3}{\partial\mathbf{n}} &\quad \text{on } \partial\omega\times(0,T),\\
\phi_3(0) = \frac{1}{2}\int_{-1}^{+1}\phi_{03}dx_3, \quad \dot{\phi}_3(0) = \frac{1}{2}\int_{-1}^{+1}\phi_{13}dx_3 &\quad \text{in } \omega,
\end{aligned}
$$

where $\{\phi_{03}, \phi_{13}\}$ *is the weak limit in* $H^1_{\Gamma_0}(\Omega)\times L^2(\Omega)$ *of the sequence* $\{\phi_{03}(\varepsilon_k), \phi_{13}(\varepsilon_k)\}_{\varepsilon_k>0}$*, with* $\{\boldsymbol{\phi}_0(\varepsilon_k), \boldsymbol{\phi}_1(\varepsilon_k)\} = \Lambda^{-1}(\varepsilon_k)(\{\mathbf{u}_1(\varepsilon_k), -\mathbf{u}_0(\varepsilon_k)\})$*. Moreover, we have*

$$
\frac{1}{2}\int_{-1}^{+1}\phi_{03}dx_3 \in H^2_0(\omega).
$$

(iii) The in-plane displacements $\hat{u}_\alpha$*,* $(\alpha = 1,2)$ *vanish in* $\omega\times(0,T)$ *(since the plate is symmetric with respect to the mid-plane* $x_3 = 0$ *and made of homogeneous material).*

Remark 6.

(a) The internal control in $(50)_1$ is such that

$$
\ddot{\phi}_3 \in [H^1(0,T;L^2(\omega))]', \quad \Delta^2\phi_3 \in L^2(0,T;H^{-2}(\omega)).
$$

By definition

$$
\langle\ddot{\phi}_3, \theta_3\rangle = -\int_{\omega\times(0,T)}\dot{\phi}_3\dot{\theta}_3 d\omega dt, \quad \forall\theta_3 \in H^1(0,T;L^2(\omega)).
$$

On the other hand

$$\langle \Delta^2\phi_3, \theta_3\rangle = \int\limits_{\omega\times(0,T)} \Delta\phi_3 \Delta\theta_3 d\omega dt, \quad \forall \theta_3 \in L^2(0,T; H_0^2(\omega)).$$

(b) In the limit problem (50) two different types of control are present. First, an internal control acting as a volume force, which roughly is the limit of the controls imposed at the upper and lower faces of the 3D plate. Second, a boundary control which is the contribution of the controls acting on the lateral boundary of the plate.

Descaled System

To better grasp the limit problem (50) it is instructive to interpret it with respect to the original plate $\bar{\Omega}^\varepsilon = \bar{\omega} \times [-\varepsilon, \varepsilon]$. To this end we define the functions w^ε, η^ε and the initial data $\{w_0^\varepsilon, w_1^\varepsilon\}$, $\{\eta_0^\varepsilon, \eta_1^\varepsilon\}$ by the *descalings* as follows, cf. Ciarlet [11],

$$\begin{aligned} w^\varepsilon(x_\alpha, t) &= \varepsilon u_3(x_\alpha, t), \quad \eta^\varepsilon(x_\alpha, t) = \varepsilon\phi_3(x_\alpha, t),\\ w_0^\varepsilon(x_\alpha) &= \varepsilon u_{03}(x_\alpha), \quad \eta_0^\varepsilon(x_\alpha) = \varepsilon\phi_{03}(x_\alpha),\\ w_1^\varepsilon(x_\alpha) &= \varepsilon u_{13}(x_\alpha), \quad \eta_1^\varepsilon(x_\alpha) = \varepsilon\phi_{13}(x_\alpha), \end{aligned}$$

for all $(x_\alpha) \in \bar{\omega}$ and all $t \in [0,T]$.

Problem (50) is then equivalent to the following one.

Theorem 5. *The descaled function w^ε is the solution (in the transposition sense) to the controlled plate problem:*

$$\begin{aligned} 2\rho^\varepsilon \varepsilon \ddot{w}^\varepsilon + \varepsilon^3 \frac{8\mu^\varepsilon(\lambda^\varepsilon + \mu^\varepsilon)}{3(\lambda^\varepsilon + 2\mu^\varepsilon)} \Delta^2 w^\varepsilon &= 2\rho^\varepsilon \varepsilon \ddot{\eta}^\varepsilon \\ + \varepsilon^2 \frac{8\mu^\varepsilon(\lambda^\varepsilon + \mu^\varepsilon)}{\lambda^\varepsilon + 2\mu^\varepsilon} \Delta^2 \eta^\varepsilon & \quad \text{in } \omega \times (0,T),\\ w^\varepsilon = 0 & \quad \text{on } \partial\omega \times (0,T),\\ \frac{\partial w^\varepsilon}{\partial \mathbf{n}} = \varepsilon^3 m_\alpha n_\alpha \Delta\eta^\varepsilon & \quad \text{on } \partial\omega \times (0,T),\\ w^\varepsilon(0) = \frac{1}{2\varepsilon}\int\limits_{-\varepsilon}^{+\varepsilon} w_0^\varepsilon dx_3^\varepsilon, \quad \dot{w}^\varepsilon(0) = \frac{1}{2\varepsilon}\int\limits_{-\varepsilon}^{+\varepsilon} w_1^\varepsilon dx_3^\varepsilon & \quad \text{in } \omega,\\ w^\varepsilon(T) = 0, \quad \dot{w}(T) = 0 & \quad \text{in } \omega. \end{aligned}$$

The control function η^ε is the solution to the system

$$\begin{aligned} 2\rho^\varepsilon \varepsilon \ddot{\eta}^\varepsilon + \varepsilon^3 \frac{8\mu^\varepsilon(\lambda^\varepsilon + \mu^\varepsilon)}{3(\lambda^\varepsilon + 2\mu^\varepsilon)} \Delta^2 \eta^\varepsilon = 0 & \quad \text{in } \omega \times (0,T),\\ \eta^\varepsilon = \frac{\partial \eta^\varepsilon}{\partial \mathbf{n}} = 0 & \quad \text{on } \partial\omega \times (0,T),\\ \eta^\varepsilon(0) = \frac{1}{2\varepsilon}\int\limits_{-\varepsilon}^{+\varepsilon} \eta_0^\varepsilon dx_3^\varepsilon, \quad \dot{\eta}^\varepsilon(0) = \frac{1}{2\varepsilon}\int\limits_{-\varepsilon}^{+\varepsilon} \eta_1^\varepsilon dx_3^\varepsilon & \quad \text{in } \omega. \end{aligned}$$

Remark 7. Similar problem of exact controllability of thin plate was considered by Saint Jean Paulin and Vanninathan [43]. In the last paper the material is still homogeneous but anisotropic cf. also [51]. Moreover, the lateral control acts only on a part of the lateral boundary and the class of exactly controllable solutions is larger.

4.3 Heat Equations in Thin Domains and Null Controllability

De Teresa and Zuazua [19] studied the problem of null controllability of linear and semilinear heat equations in thin domain $\Omega^\varepsilon = \Omega \times (0, \varepsilon)$ where Ω is a bounded open smooth domain of $\mathbb{R}^{n-1}$ ($n \geq 2$). By ω we now denote a non-empty subset of Ω; $\omega^\varepsilon = \omega \times (0, \varepsilon)$ stands for corresponding cylindrical subset of Ω^ε.

Given a control time $T > 0$ and a potential $a^\varepsilon(\mathbf{x}', x_n, t) \in L^\infty(Q^\varepsilon)$, $Q^\varepsilon = \Omega^\varepsilon \times (0, T)$, the following heat equation was considered in [19]:

$$\begin{aligned} \dot{u}_t^\varepsilon - \Delta u^\varepsilon + a^\varepsilon(\mathbf{x}', x_n, t) u^\varepsilon = h^\varepsilon \chi_{\omega^\varepsilon} \quad & \text{in } Q^\varepsilon, \\ u^\varepsilon = 0 \quad & \text{on } [\partial\Omega \times (0, \varepsilon)] \times (0, T), \\ \frac{\partial u^\varepsilon}{\partial \mathbf{n}} = 0 \quad & \text{on } \Gamma_\pm^\varepsilon \times (0, T), \\ u^\varepsilon(0) = u_0^\varepsilon \quad & \text{in } \Omega^\varepsilon. \end{aligned} \tag{51}$$

Here $\mathbf{x} = (\mathbf{x}', x_n)$, $h = h(\mathbf{x}, t)$ is the control function, and $\chi_{\Omega^\varepsilon}$ denotes the characteristic function of ω^ε. System (51) is known to be *null controllable* in *any* time $T > 0$, i.e. there exists a control $h^\varepsilon \in L^2(\omega^\varepsilon \times (0, T))$ such the solution of (51) satisfies $u^\varepsilon(0) = 0$ in Ω^ε.

Similarly to Sec. 4.2, the variable x_n is rescaled in order to work in a reference domain $\Omega^1 = \Omega \times (0, 1)$. The following change of variables is made:

$$\begin{aligned} & x_n = \varepsilon z_n, \quad u(\varepsilon)(\mathbf{x}', z_n, t) = u(\mathbf{x}', \varepsilon z_n, t), \\ & g(\varepsilon)(\mathbf{x}', z_n, t) = h(\mathbf{x}', \varepsilon z_n, t), \quad u_0(\varepsilon)(\mathbf{x}', z_n) = u_0^\varepsilon(\mathbf{x}', \varepsilon z_n), \\ & b(\varepsilon)(\mathbf{x}', z_n, t) = a(\mathbf{x}', \varepsilon z_n, t). \end{aligned}$$

The proof of null controllability of system (51) exploits in an essential manner its parabolic nature. The approach used in [19] reduces to obtaining a suitable observability inequality for the system adjoint to (51). That inequality relies on *Carleman inequalities.* In [19] the authors extended some results on Carleman estimates due to Fursikov and Imanuvilov [23], see also Tataru [46, 47], Albano [1], and Albano and Tataru [2].

To get a flavour of such estimates consider the following simpler heat equation [23]:

$$\begin{aligned} & \dot{\theta} + \Delta\theta = f(\mathbf{x}, t) \quad \text{in } \tilde{Q} = \tilde{\Omega} \times (0, T), \\ & \theta = 0, \quad \frac{\partial \theta}{\partial \mathbf{n}} = 0 \quad \text{on } \partial\tilde{\Omega} \times (0, T). \end{aligned} \tag{52}$$

Here $\tilde{\Omega} \subset \mathbb{R}^n$ is a connected bounded domain of class C^2. Then the following results hold true.

Lemma 2. *There exists $s_0 > 0$ such that for any $s > s_0$ the solution $\theta(\mathbf{x}, t)$ to (52) satisfies the Carleman inequality:*

$$\int_{\tilde{Q}} \left[(s\varphi)^{-1} \left((\dot{\theta})^2 + \sum_{i,j=1}^{n} \left| \frac{\partial^2 \theta}{\partial x_i \partial x_j} \right|^2 \right) + s\varphi \sum_{i=1}^{n} \left| \frac{\partial \theta}{\partial x_i} \right|^2 + s^3 \varphi^3 \theta^2 \right]$$
$$\times e^{s\alpha(\mathbf{x},t)} d\mathbf{x}dt \leq c \int_{\tilde{Q}} f^2(\mathbf{x}, t) e^{s\alpha} d\mathbf{x}dt,$$

where $c > 0$ does not depend on s and

$$\varphi(\mathbf{x}, t) = e^{\Psi(\mathbf{x})}/t(T-t),$$
$$\alpha(\mathbf{x}, t) = \left(e^{\Psi(\mathbf{x})} - e^{2\|\Psi\|_{C(\bar{\Omega})}} \right) /(t(T-t)).$$

5 Controllability and Homogenisation

This type of studies was initiated, as far as we know, by Lions [34]. In [34] linear evolution operator of hyperbolic type with microperiodic coefficients and internal control was investigated. Combining homogenisation with HUM the problem of exact controllability of the limit problem as $\varepsilon \to 0$ was solved. As usual in the homogenisation, ε stands for a small positive parameter, intended to tend to zero, characterizing the microstructure. This result was next extended by Telega and Bielski [52] to internal controllability of the equations of dynamic elasticity. Lions [34] studied also the case of boundary controllability, but only formally (in the case of scalar equation of hyperbolic type). The case of boundary controllability is much more complex as evidenced by a one-dimensional counterexample considered by Avellaneda et al. cited in [5, 6]. These authors proved that there exist stationary solutions which concentrate most of their energy in one part of the boundary. These solutions contain frequencies of the order of the inverse of the oscillation parameter ($\sim \varepsilon^{-1}$), i.e. the solution with wavelength of the order of ε. This type of phenomena constitutes an obstacle to the uniform boundary controllability when the microstructure becomes finer and finer.

5.1 Wave Equations

To overcome the difficulty mentioned, Castro [5] proposed to remove from the solution the Fourier components corresponding to the frequencies with wavelength of the order of ε. More precisely, this author considered the following one-dimensional wave equation:

$$
\begin{aligned}
&\rho^\varepsilon \ddot{u} - u_{xx} = 0, && 0 < x < 1, && 0 < t < T,\\
&u(0,t) = 0, && u(1,t) = f(t), && 0 < t < T,\\
&u(x,0) = u_0(x), && \dot{u}(x,0) = u_1(x), && 0 < x < 1,
\end{aligned}
\tag{53}
$$

where $f \in L^2(0,T)$ is a *boundary control function* acting on the extreme point $x = 1$ of the vibrating string. As usual, (u_0, u_1) are the initial data. It is assumed that $\rho \in L^\infty(\mathbb{R})$ is a periodic function such that $0 < \rho_m < \rho(x) < \rho_M$ a.e. $x \in \mathbb{R}$. As usual, we set $\rho^\varepsilon(x) = \rho(x/\varepsilon)$, cf. [31, 44]. We recall that $\ddot{u} = \frac{\partial^2 u}{\partial t^2}$, $u_{xx} = \frac{\partial^2 u}{\partial x^2}$; moreover $\varepsilon = l/L$ where l denotes the typical length of the microstructure and L is the typical macroscopic length (in this specific case $L = 1$).

Castro [5] proved that for system (53) the following statements hold:

(i) If $\rho^\varepsilon \in W^{1,\infty}(0,1)$, $(u_0,u_1) \in L^2(0,1) \times H^{-1}(0,1)$, $f \in L^2(0,T)$ then $(u^\varepsilon, \dot{u}^\varepsilon)$, the solution to (53), blows up as $\varepsilon \to 0$ since in the upper bound the term $\|\rho^\varepsilon\|^{1/2}_{W^{1,\infty}(0,1)}$ is present.

(ii) If $\rho^\varepsilon \in BV(0,1)$ (the space of functions with bounded variation) then f^ε (the control function for $\varepsilon > 0$) blows up exponentially as $\varepsilon \to 0$. For $0 < \varepsilon < 1$ and $T > 2\sqrt{\rho_M}$, system (53) is null controllable, i.e. $u^\varepsilon(T) = 0$, $\dot{u}^\varepsilon(T) = 0$. We observe that $\|\rho^\varepsilon\|_{W^{1,\infty}(0,1)} \sim \varepsilon^{-1}$.

Remark 8. There exist many possible controls $f^\varepsilon \in L^2(0,T)$ driving system (53) to rest. The advantage of HUM is that it yields the control function with minimal L^2-norm.

It is well-known that $\rho^\varepsilon(x) = \rho(x/\varepsilon)$ converges weakly to the average value $\bar{\rho} = \int_0^1 \rho(s)ds$ as $\varepsilon \to 0$, cf. [31, 34]. Moreover, (53) H-converges to the limit system

$$
\begin{aligned}
&\bar{\rho}\ddot{u} - u_{xx} = 0, && 0 < x < 1, && 0 < t < T,\\
&u(0,t) = 0, && u(1,t) = f(t), && 0 < t < T,\\
&u(x,0) = u_0(x), && \dot{u}(x,0) = u_1(x), && 0 < x < 1.
\end{aligned}
\tag{54}
$$

However, one cannot establish estimates for $\{u^\varepsilon, \dot{u}^\varepsilon\}_{\varepsilon>0}$ and $\{f^\varepsilon\}$ uniform in ε. This can be done provided that one removes from u^ε, the solution of (53), the Fourier components corresponding to the frequencies with wavelength of the order of ε. Such a statement holds true when $\rho \in L^\infty\mathbb{R}$. In the case of high frequencies, more and more regularity of ρ is needed to cover the whole range of frequencies with wavelength greater than ε. Detailed mathematical analysis was performed by Castro [5] and Castro and Zuazua [6, 7]. The methods used comprise spectral analysis, Ingham inequality, HUM and WKB theory (Wentzel-Kramers-Brillouin theory, cf. [4, Chap. 10]. Unfortunately, the approach used is restricted to the one-dimensional case.

Remark 9. Lebeau [30] extended the observability estimate due to Castro and Zuazua [6] to the multidimensional case (observability at low frequency).

Remark 10. Castro and Zuazua [8], showed that the assumption $\rho \in BV(0,1)$ mentioned earlier is essential. More precisely, consider the following one-dimensional wave equation:

$$\begin{aligned} \rho(x)\ddot{v} - v_{xx} = 0, \quad & 0 < x < 1, \quad 0 < t < T, \\ v(0,t) = v(1,t) = 0, \quad & 0 < t < T, \\ v(x,0) = v_0(x), \quad \dot{v}(x,0) = v_1(x), \quad & 0 < x < 1. \end{aligned} \tag{55}$$

We assume that ρ is measurable and

$$\exists \rho_M \geq \rho_m > 0, \quad 0 < \rho_m \leq \rho(x) \leq \rho_M < +\infty \quad \text{a.e. } x \in (0,1). \tag{56}$$

Then system (55) is well-posed in the sense that for any pair of initial data $(v_0, v_1) \in H_0^1(0,1) \times L^2(0,1)$ there exists a unique solution

$$v \in C([0,T]; H_0^1(0,1)) \cap C^1([0,T]; L^2(0,1)).$$

Since the system is conservative, the energy of solutions

$$E(t) = \frac{1}{2} \int_0^1 \left[\rho(x)|\dot{v}(x,t)|^2 + |v_x(x,t)|^2\right] dx \tag{57}$$

is constant in time, i.e. $E(0) = E(t)$. If $\rho \in BV(0,1)$, the following observability properties are well established:

(i) **Boundary observability:** if $T > \sqrt{\rho}_M$ there exists $C(T) > 0$ such

$$E(0) \leq C \int_0^T \left[|v_x(0,t)|^2 + |v_x(1,t)^2|\right] dt, \tag{58}$$

for every solution of (55).

(ii) **Internal observability:**
For any subinterval $(\alpha, \beta) \subset (0,1)$, if $T > 2\sqrt{\rho}_M \max(\alpha, 1-\beta)$, there exists $C_1 > 0$ such that

$$E(0) \leq C_1 \int_0^T \int_\alpha^\beta \left[\rho(x)\dot{v}^2 + v_x^2\right] dx dt, \tag{59}$$

for every solution of (55).

We observe that these observability estimates are equivalent to the controllability of the system with controls acting on the boundary or in the interior of the domain, respectively.

According to Castro and Zuazua [8] for a long time the problem whether the estimates (58), (59) hold for less regular coefficients like $\rho \in L^\infty(0,1)$ or $\rho \in C([0,1])$ has been left open. We already know that the observability inequalities (58), (59) blow up when ρ is replaced by $\rho^\varepsilon = \rho(x/\varepsilon)$ and $\varepsilon \to 0$. Castro and Zuazua [8], proved, in a constructive manner, the following result.

Theorem 6. *There exists Hölder continuous density function $\rho \in C^{0,s}([0,1])$ for all $0 < s < 1$, for which (58) and (59) fail for all $T > 0$ and for all subinterval $(\alpha, \beta) \subset (0,1)$ where $(\alpha,\beta) \neq (0,1)$.*

The proof of Th. 6 is based on a construction of density ρ for which there exists a sequence of pairs $(\varphi_k(x), \lambda_k)$ satisfying

$$\frac{d^2\varphi_k}{dx^2} + \lambda_k^2 \rho(x)\varphi_k = 0,$$

and such that φ_k is exponentially concentrated on any given point of the closed interval $[0,1]$.

Remark 11. The last theorem can be extended to higher dimensional wave equation [8]. Indeed, by exploiting the 1D approach one can construct densities ρ in separated variables, which oscillate in a neighbourhood of any point of the domain or of the boundary. For details the reader is referred to [8].

5.2 Controllability of Parabolic Equations with Oscillating Coefficients

In this section we shall synthesize the available results on controllability of heat equations with oscillating coefficients in one- and multi- dimensional cases.

One-Dimensional Case

Let the density $\rho \in W^{2,\infty}(\mathbb{R})$ be a periodic function such that (56) is satisfied for all $x \in \mathbb{R}$. As usual, $W^{2,\infty}(\mathbb{R})$ is the Sobolev space of $L^\infty(\mathbb{R})$ functions with the first and the second derivatives in $L^\infty(\mathbb{R})$.

Consider the following heat equation with microperiodic density:

$$\begin{aligned} \rho\left(\frac{x}{\varepsilon}\right)\dot{u}^\varepsilon - u^\varepsilon_{xx} = 0, \quad 0 < x < 1, \quad 0 < t < T, \\ u^\varepsilon(0,t) = 0, \quad u^\varepsilon(1,t) = f^\varepsilon(t), \quad 0 < t < T, \\ u^\varepsilon(x,0) = u_0(x), \quad 0 < x < 1. \end{aligned} \tag{60}$$

Classical results show that for any $T > 0$, $\varepsilon \in (0,1)$ and $u_0 \in L^2(0,1)$ there exists a control $f^\varepsilon \in L^2(0,T)$ such that the solution to (60) satisfies

$$u^\varepsilon(x,T) = 0, \quad \text{for all } 0 < x < 1.$$

Indeed, for all $T > 0$ and $\varepsilon \in (0,1)$ there exists a positive constant $C(\varepsilon,T)$ such that

$$\|f^\varepsilon\|_{L^2(0,T)} \leq C(\varepsilon,T)\|u_0\|_{L^2(0,1} \tag{61}$$

for all $u_0 \in L^2(0,1)$. Tha main difficulty consists in showing that $C(\varepsilon,T)$ remains bounded as $\varepsilon \to 0$.

The formal limit of (60) as $\varepsilon \to 0$ is the averaged system:

$$\begin{aligned} \bar{\rho}\dot{u} - u_{xx} = 0, \quad 0 < x < 1, \quad 0 < t < T, \\ u(0,t) = 0, \quad u(1,t) = f(t), \quad 0 < t < T, \\ u(x,0) = u_0(x), \quad 0 < x < 1. \end{aligned} \tag{62}$$

López and Zuazua [36] proved that the constant C in (61) does not depend on ε and for any $u_0 \in L^2(0,1)$ fixed, there exists a sequence of controls $\{f^\varepsilon\}_{\varepsilon>0}$ for system (60) such that

$$f^\varepsilon \to f \text{ strongly in } L^2(0,T) \text{ as } \varepsilon \to 0,$$

f being a control of the limit system (62), so that the solution to (62) satisfies

$$u(x,T) = 0, \qquad \text{for all } 0 < x < 1.$$

The sequence $\{u^\varepsilon\}_{\varepsilon>0}$ of solutions to (60) is bounded in $L^\infty(0,T;H^{-1}(0,1))$ and $u^\varepsilon \rightharpoonup u$ weakly-$\star$ in $L^\infty(0,T;H^{-1}(0,1))$ as $\varepsilon \to 0$ for all $s > \frac{3}{2}$.

Grosso modo, the proof runs as follows. We divide the time interval $[0,T]$ in three subintervals: $I_1 = [0,T/3]$, $I_2 = [T/3,2T/3]$, $I_3 = [2T/3,T]$. In I_1 we control to zero the projection of the solution over a suitable subspace containing only sufficiently *low frequencies.* In I_2 we let the system to evolve freely without control. In this way, the projection of solution to (60) over low frequencies remains at rest and, due to the strong dissipativity of (60) in its *high frequencies*, the size of the solution at time $t = 2T/3$ becomes exponentially small, of the order of $\exp(-c/\varepsilon^2)$ for a suitable constant $c > 0$, as $\varepsilon \to 0$. Finally, in I_3 we apply a control driving the whole solution to zero. Using Carleman estimates it can be shown that the control needed in the third interval is at most of order $\exp(c_1/\varepsilon^{4/3})$ for a suitable constant $c_1 > 0$, with respect to the solution at time $t = 2T/3$. However, since, in view of the analysis performed in I_2, the norm of the solution, in the absence of control, at $t = 2T/3$ decays by a multiplicative factor of the order of $\exp(-c_1/\varepsilon^2)$, these two phenomena compensate and the control needed in I_3 turns out to be uniformly bounded, and even exponentially small as $\varepsilon \to 0$. Thus we conclude that the dissipativity of the heat equation compensates the spectral pathologies leading to the blow-up of the observability constant in the context of the wave equation, cf. Sec. 5.1 of our paper.

Remark 12.

(i) The control strategy in three steps can also be applied to the case $\rho \in W^{1,\infty}(\mathbb{R})$. In contrast, the case $\rho \in L^\infty(\mathbb{R})$ remains an open problem [9].

(ii) The proof of the controllability results of system (60) can be extended to the following system [9]:

$$\begin{aligned} \dot{u}^\varepsilon - (a(x/\varepsilon)u^\varepsilon_x)_x = 0, \quad 0 < x < 1, \quad 0 < t < T, \\ u^\varepsilon(0,t) = 0, \quad u^\varepsilon(1,t) = f^\varepsilon(t), \quad 0 < t < T, \\ u^\varepsilon(x,0) = u_0(x), \quad 0 < x < 1, \end{aligned}$$

where $a \in W^{2,\infty}(\mathbb{R})$ is a periodic function bounded from above and from below. Now the limit system involves the *homogenized coefficient* a^h where

$$a^h = \left(\int_0^1 \frac{ds}{a(s)} \right)^{-1}. \tag{63}$$

For details on the derivation of (63) the reader is referred to [44].

(iii) Internal approximate controllability of the n-dimensional heat equation with rapidly oscillating diffusion coefficients was considered by Zuazua [58], cf. also [9].

(iv) Castro and Zuazua [10] analysed the approximate controllability of the following system

$$\begin{aligned} \dot{u} - \Delta u = f(\mathbf{x},t)\delta_{\gamma(t/\varepsilon)}(\mathbf{x}) \quad & \text{in } Q = \Omega \times (0,T), \\ u = 0 \quad & \text{on } \Sigma = \partial\Omega \times (0,T), \\ u(\mathbf{x},0) = u_0(\mathbf{x}) \quad & \text{in } \Omega. \end{aligned}$$

Here Ω is a bounded smooth domain in $\mathbb{R}^n$ ($n = 1,2,3$) and f is a control function; δ_γ represents the Dirac measure on γ. Here γ can be, for instance, a point, a curve if $n \geq 2$, or a surface if $n = 3$. In [10] the case where $\{\gamma(t)\}_{t \leq T}$ is a non-constant periodic family of $k = (n-1)$-dimensional manifolds was considered ($\varepsilon > 0$ is a small parameter). For $\varepsilon \to 0$ the oscillating controllers undergo a homogenisation process. In the limit we get controls distributed on a time-independent manifold of dimension $k + 1$ modulated by a density factor varying along the manifold. Hence we conclude that time oscillations on the support of the control enhance the controllability properties.

(v) Castro and Zuazua [9] considered the finite difference space semi-discretization of the classical 1D heat equation (no oscillations). It was shown that there is a strong analogy between homogenisation and numerical approximation since in the latter case the small parameter h of the mesh-size plays the role of the ε describing the scale of oscillations or heterogeneities of the medium, cf. also [56, 57].

5.3 Internal and Boundary Controllability of Wave Equation in Perforated Domains

In 1988 Lions introduced a new class of controllability problems, namely those in perforated domains, cf. [34, Chap. 5]. This author studied exact internal controllability in the different cases of Neumann and Dirichlet homogeneous conditions imposed on the boundaries of *periodically* distributed microholes of the size of a small parameter $\varepsilon > 0$. Combining HUM with the homogenisation method of the two scale asymptotic developments a limit problem was derived. For more detail on homogenisation in perforated domains the reader is referred to [18, 27, 31] and the relevant references cited therein. Rigorous proofs in the case of the homogeneous Neumann condition imposed on the boundaries of holes were carried out by Cioranescu and Donato [12].

Saint Jean Paulin and Tcheugoué Tébou [42] studied more general case of exact internal controllability in the case with a specific Fourier type condition imposed on the boundaries of microholes. More precisely, let Ω be a bounded and regular domain in $\mathbb{R}^n$, $n \geq 2$. The domain Ω is periodically perforated with holes on the order of ε with period εY where $Y = (0,1)^n$ is the so-called basic cell [27, 31, 44]. Let S be an open subset of Y with the boundary of class C^2. The set S_ε of micropores contained in Ω is defined by

$$S_\varepsilon = \Omega \cap \left(\bigcup_{\mathbf{k} \in \mathbb{Z}^n} \varepsilon(S + \mathbf{k}) \right);$$

i.e. the micropores are constructed by translation and homothety of S; $\mathbb{Z}$ denotes the set of integers. It is assumed that the micropores do not intersect the boundary of Ω. We set $Y^\star = Y \setminus \bar{S}$, $\Omega_\varepsilon = \Omega \setminus \bar{S}_\varepsilon$. Consider the following internal controllability problem:

$$\begin{aligned} \ddot{u}_\varepsilon + A_\varepsilon u_\varepsilon = v_\varepsilon \quad & \text{in } \Omega_\varepsilon \times (0,T), \\ \frac{\partial u_\varepsilon}{\partial \mathbf{n}_{A^\varepsilon}} + \alpha \varepsilon^\mu u_\varepsilon = 0 \quad & \text{on } \partial S_\varepsilon \times (0,T), \\ u_\varepsilon(0) = u_\varepsilon^0, \quad \dot{u}_\varepsilon(0) = u_\varepsilon^1 \quad & \text{in } \Omega_\varepsilon. \end{aligned} \tag{64}$$

Here α and μ are real numbers with $\mu \geq 1$ and

$$A_\varepsilon = -\frac{\partial}{\partial x_i}\left(a_{ij}\left(\frac{\mathbf{x}}{\varepsilon}\right) \frac{\partial}{\partial x_j} \right), \quad \frac{\partial}{\partial \mathbf{n}_{A_\varepsilon}} = a_{ij}\left(\frac{\mathbf{x}}{\varepsilon}\right)\frac{\partial}{\partial x_j} n_i.$$

We recall that $\frac{\partial}{\partial \mathbf{n}_{A_\varepsilon}}$ is the conormal derivative of the operator A_ε. It is assumed that

$$a_{ij} \in L^\infty(\mathbb{R}^n), \quad a_{ij} \text{ is } Y\text{-periodic}, \ a_{ij} = a_{ij},$$

$$\exists c_0 > 0 : a_{ij}(\mathbf{y})\xi_i\xi_j \geq c_0 \xi_i \xi_j \quad \forall \boldsymbol{\xi} \in \mathbb{R}^n, \text{ a.e. } \mathbf{y} \in \mathbb{R}^n.$$

Two problems were solved in [42]: (i) null-controllability of (64), and (ii) the limit behaviour of u_ε and v_ε (control function when $\varepsilon \to 0$). It is not difficult to solve the first problem when $\varepsilon > 0$ is fixed. To this end it suffices to use the Hilbert Uniqueness Method. The passage to the limit is more complicated and combines homogenisation of perforated domains with HUM. Such a homogenisation involves extension operators [27, 31].

Denote by $\tilde{f}$ an extension of a function f defined in Ω_ε. Under suitable, physically plausible assumptions on the initial values u_ε^0 and u_ε^1 we have

$$\tilde{v}_\varepsilon \rightharpoonup v \text{ weakly in } L^2(0,T;L^2(\Omega))$$

where v is the control function of the homogenised system:

$$\begin{aligned} \theta u + A^h u + C(\theta,\alpha,\mu) &= v, && \text{in } \Omega \times (0,T), \\ u &= 0, && \text{on } \partial\Omega \times (0,T), \\ u(0) &= u^0/\theta, && \text{in } \Omega, \\ \dot{u}(0) &= u^1/\theta, && \text{in } \Omega. \end{aligned}$$

Here $\theta = |Y^\star|/Y$, $A^h = -a_{ij}^h \frac{\partial^2}{\partial x_i \partial x_j}$, a_{ij}^h are homogenised moduli [27, 44] and

$$C(\theta,\alpha,\mu) = \begin{cases} \theta\alpha|\partial S|/|Y^\star| & \text{if } \mu = 1, \\ 0 & \text{if } \mu > 0. \end{cases}$$

The (weak-$\star$) convergence of u_ε and $\dot{u}_\varepsilon$ involves an extension operator P_ε, say from Ω_ε to Ω.

6 Final Remarks

In a series of papers [14–17], the authors studied some boundary controllability problems in domains with small holes, smaller than those considered in [34, 42, 52]. Then, by using the homogenisation results for perforated domains [13], with not necessarily periodically distributed micropores, a "strange term" in the wave equation appears in the case of the "critical size" of the pores.

Tcheugoué Tébou [48] investigated stabilisation and internal controllability via Russell's principle [41] of grids involving *three small parameters.*

Hoffmann and Botkin [26] studied formally the problem of controllability of von Kármán plates with periodically distributed piezoelectric patches, cf. also [24, 25]. The model is not very realistic since the piezoelectric actuators are modelled as acting on the mid-plane of the plate. Physically, they should be distributed on the faces of the plate [3] or shell [35].

As one can infer from our presentation, controllability and stabilisation results involving a small parameter intended to tend to zero are in fact restricted to more or less model problems. Many open problems arise. For instance, problems involving nonlinear homogenisation of deformable solids and structures combined with controllability seem to be untouched except the paper [26].

Acknowledgement. The author is indebted to Mr Maciej Stańczyk for his help in typesetting of the manuscript. The paper was partially supported by the Ministry of Science and Information Technology (Poland) through the grant No8 T07A 052 21.

References

1. Albano P (2000) Carleman estimates for the Euler-Bernoulli plate operator. Electronic J. Diff. Eqs. 2000(53):1–13
2. Albano P, Tataru D (2000) Carleman estimates and boundary observability for a coupled parabolic-hyperbolic system. Electronic J. Diff. Eqs. 2000(22):1–15
3. Banks HT, Smith RC, Wang Y (1996) Smart materials structures: modeling, estimation and control. Wiley, Chichester
4. Bender CM, Orszag SA (1978) Advanced mathematical methods for scientists and engineers. McGraw-Hill, New York
5. Castro C (1999) Boundary controllability of the one-dimensional wave equation with rapidly oscillating density. Asymptotic Anal. 20:317–350
6. Castro C, Zuazua E (2000) Low frequency asymptotic analysis of a string with rapidly oscillating density. SIAM J. Appl. Math. 60:1205–1233
7. Castro C, Zuazua E (2000) High frequency asymptotic analysis of a string with rapidly oscillating density. Euro. J. Appl. Math. 11:595–622
8. Castro C, Zuazua E (2003) Concentration and lack of observability of waves in highly heterogeneous media. Arch. Rat. Mech. Anal. 164:39–72
9. Castro C, Zuazua E Some topics on the control and homogenisation of parabolic partial differential equations. in press
10. Castro C, Zuazua E Unique continuation and control for the heat equation from an oscillating lower dimensional manifold. in press
11. Ciarlet PG (1997) Mathematical elasticity, vol. II: Theory of plates. North-Holland, Amsterdam
12. Cioranescu D, Donato P (1989) Exact internal controllability in perforated domains. J. Math. Pures. Appl. 68:185–213
13. Cioranescu D, Murat F (1982) Un terme étrange venu d'ailleurs. In: Brezis H, Lions J-L, Cioranescu D (eds) Nonlinear partial differential equations and their applications. Collége de France Seminar vol. 2(6):98–138, vol. 3:154–178, Research Notes in Mathematics, Pitman, London, also in: Cherkaev A, Kohn R (eds) Topics in the Mathematical Modelling of Composite Materials, pp. 45–93, Birkhäuser, Boston (1997)
14. Cioranescu D, Donato P (1989) Contrôlabilité exacte frontiére de l'équation des ondes dans des domaines avec des petits trous. C.R. Acad. Sci. Paris, Série I 309:473–478
15. Cioranescu D, Donato P, Zuazua E (1991) Strong convergence on the exact boundary controllability of the wave equation in the perforated domains. In: Desch W, Kappel F, Kunisch K (eds) Proc. Int. Conf. on Control and Estimation of Distributed Parameter Systems, Vorau (Styria), ISNM Series, vol. 100:99-113, Birkhäuser Verlag, Basel
16. Cioranescu D, Donato P, Zuazua E (1992) Exact boundary controllability for the wave equation in domains with small holes. J. Math. Pures Appl. 71:343–377
17. Cioranescu D, Donato P, Zuazua E (1994) Approximate boundary controllability for the wave equation in perforated domains. SIAM J. Control Optim. 32:35–50
18. Cioranescu D, Donato P, Murat F, Zuazua E (1991) Homogenization and correctors for the wave equation in domains with small holes. Ann. Scuola Norm. Sup. Pisa 18:251–293

19. de Teresa L, Zuazua E (2000) Null controllability of linear and semilinear heat equations in thin domains. Asymptotic Anal. 24:295-317
20. Fabre C (1990) Comportement au voisinage du bord des solutions de quelques équations d'évolutions linéaires. Application á certains problémes de contrôlabilité exacte et de perturbations singuliéres. Thése de doctorat, Université Paris VI
21. Figueiredo I, Zuazua E (1995) Exact controllability, asymptotic methods and plates. In: Ciarlet PG, Trabucho, Viano (eds) Asumptotic methods for elastic structures, pp.69–74, Walter de Gruyter&Co., Berlin-New York
22. Figueiredo I, Zuazua (1996) Exact controllability and asymptotic limit for thin plates. Asymptotic Anal. 12:213-252
23. Fursikov AV, Imanuvilov D Yu (1996) Controllability of evolution equations. Research Institute of mathematics, Global Analysis Research Center, Seoul National University, Seoul
24. Hoffmann K-H Botkin ND (1998) Oscillation of nonlinear thin plates excited by piezoelectric patches. Z. Angew. Math. Mech. 78:495–503
25. Hoffmann K-H Botkin ND (1999) A fully coupled model of a nonlinear thin plate excited by piezoelectric actuator. In: Hoffmann K-H (ed) Proceedings of the first caesarium, Springer, available via the preprint server: `http://www.zib.de/dfg-echtzeit/Publikationen/index.html`
26. Hoffmann K-H Botkin ND (2000) Homogenization of von Kármán plates excited by piezoelectric patches. Z. Angew. Math. Mech. 80:579–590
27. Jikov VV, Kozlov SM, Oleinik OA (1994) Homogenization of differential operators and integral functionals, Springer-Verlag, Berlin
28. Komornik V (1994) Controllability and stabilization: the multiplier method. RAM36, John Wiley&Sons, Chichester; Masson, Paris
29. Lasiecka I, Triggiani R (2000) Control theory for partial differential equations: continuous and approximate theories, I. Abstract parabolic systems, II. Abstract hyperbolic-like systems over a finite time horizon. Cambridge University Press, Cambridge
30. Lebeau G (2000) The wave equation with oscillating density: observability at low frequency. ESAIM, COCV, 5:219–258
31. Lewiński T, Telega JJ (2000) Plates, laminates and shells: asymptotic analysis and homogenisation. World Scientific Publishers, Singapore
32. Lions JL (1985) Asymptotic problems in distributed systems. IMA Preprint Series No 147, May 1985, Univeristy of Minnesota
33. Lions J-L (1988) Contrôlabilité exacte, perturbations et stabilisation de systémes distribués. Vol. 1, Contrôlabilité Exacte, RMA8, Masson, Paris
34. Lions J-L (1988) Contrôlabilité exacte, perturbations et stabilisation de systémes distribués. Vol. 2, Perturbations, RMA9, Masson, Paris
35. Litvinov WG, Kröplin B Modeling and optimal control of coupled structural acoustic systems with piezoelectric elements. Math. Meth. Appl. Math. in press
36. López A, Zuazua E Uniform null-controllability for the one-dimensional heat equation with rapidly oscillating periodic density. in press
37. López A, Zhang X, Zuazua E (2000) Null controllability of the heat equation as singular limit of the exact controllability of dissipative wave equations. J. Math. Pures. Appl. 79:741–808
38. Micu JS, Zuazua E (2004) An introduction to the controllability of partial differential equations. in press

39. Rodriguez-Bernal A, Zuazua E (1995) Parabolic singular limit of a wave equation with localized boundary damping. Discrete Cont. Dyn. Systems 1:303–346
40. Rodriguez-Bernal A, Zuazua E (2001) Parabolic singular limit of a wave equation with localized interior damping. Commun. Contemporary Math.3:215–257
41. Russell DL (1978) Controllability and stabilizability theory for linear partial differential equations. Recent progress and open questions. SIAM Review 20:639–739
42. Saint Jean Paulin J, Tcheugoué Tébou LR (1997) Contrôlabilité exacte interne dans des domaines perforés avec une condition aux limite de Fourier sur le bord des trous. Asymptotic Anal. 14:193–221
43. Saint Jean Paulin J, Vanninathan M (1996) Contrôlabilité exacte de vibrations de corps élastiques minces. C.R. Acad. Sci. Paris, Série I 322:889–894
44. Sanchez-Palencia E (1980) Non-homogeneous media and vibration theory. Lecture Notes in physics 127, Springer-Verlag, Berlin
45. Sławianowska A, Telega JJ (2001) A dynamic asymptotic model of linear elastic orthotropic plates: first and second-order terms. Arch. Mech. 53:541–564
46. Tataru D (1994) A priori estimates of Carleman's type in domains with boundary. J. Math. Pures Appl. 73:355–387
47. Tataru D (1996) Carleman estimates and unique continuation for solutions to boundary value problems. J. Math. Pures Appl. 75:367–408
48. Tcheugoué Tébou LR (1996) Internal stabilization and exact controllability in thin cellular structures. Ric. Mat. 45:457–490
49. Tcheugoué Tébou LR (1997) Contrôle distribué de l'equation des ondes dans des domaines minces. Math. Modelling Numer. Anal. 31:871–890
50. Tcheugoué Tébou LR (1998) Sur quelques résultats d'observabilité liés á l'équation des ondes perturbée. C.R. Acad. Sci. Paris, Série I 327:277–281
51. Telega JJ, Bielski W (1996) Exact controllability of anisotropic elastic bodies. In: Malanowski K, Nahorski Z, Peszyńska M (eds) Modelling and optimization of distributed parameter systems. pp.254–262, Chapman Hall, London
52. Telega JJ, Bielski W (1999) On two problems of exact controllability for anisotropic elastic solids. In: Holnicki-Szulc J, Rodellar J (eds) Smart Structures. pp. 345–354, Kluwer Academic Publishers, Dordrecht
53. Telega JJ, Bielski W (2000) Controllability and stabilization in elasticity, heat conduction and thermoelasticity: review of recent developments. J. Global Optim. 17:353–386
54. Telega JJ (2001) Topics on deterministic and stochastic controllability and stabilization of parameter distributed systems: theory and numerical approximations. In: Holnicki-Szulc J (ed) Structural control and health monitoring. pp. 213–340, AMAS Lecture Notes, vol. 1, IFTR Warsaw
55. Yan J (1992) Contrôlabilité exacte pour des domaines minces. Asymptotic Anal. 5:461–471
56. Zuazua E (2002) Propagation, observation, control and numerical approximation of waves. preprint, Departmendo de Matemáticas, Universidad Autónoma, Madrid, Spain
57. Zuazua E Controllability of partial differential equations and its semi-discrete approximations. in press
58. Zuazua E (1994) Approximate controllability for linear parabolic equations with rapidly oscillating coefficients. Control and Cybernetics 23:793–808

Mechatronics in Vibration Monitoring and Control

Tadeusz Uhl

University of Science and Technology AGH,
al. Mickiewicza 30, 30-059 Kraków, Poland
tuhl@agh.edu.pl

Summary. This chapter deals with methods of vibration monitoring and control based on mechatronic solutions. An innovative, new approach of design and implementation of mechatronic systems is applied to design of smart sensor for operational load measurements. New idea of smart sensors and their application for health monitoring and diagnostics of structures is described. Test of performance of designed and implemented smart sensor is shown. The main advantage of its application is automation of design process focused on virtual and fast prototyping. The second example of application of mechatronic design method is active balancing system for rotating shaft which is carefully discussed. The concept, control algorithms, and prototyping system implementation are shown.

1 Introduction

Next generation systems such as aircraft, spacecraft, missiles, high speed trains, submarines and others will be required to achieve the desired levels of performance in many different operating regimes and under different failures and upsets. These requirements create new complex tasks for on-line health monitoring systems which will be in the future standard part of design. Nowadays, Health Monitoring and Failure Detection, Identification and Reconfiguration (HM-FDIR) systems have been widely applied in different areas [1]. Many different techniques have been developed and successfully implemented. One possibility of health monitoring system realization is to estimate their dynamic properties, mainly based on vibration phenomena. But, from other side, currently designed structures are more and more complex and require more effective tools for measurements, vibration control and diagnostics. One possibly way to solve very complex HM-FDIR system design problems is mechatronic approach.

Mechatronics is synergetic combination of mechanical engineering, electronic, control and computer science in design of products and processes [2]. Mechatronic systems are interdisciplinary systems and to achieve their performance enough good, all components should be treated during design and prototyping with same weight. To achieve defined above objectives, design process should integrate all fields, which are included in design system. It means, parts of different nature should co-operate in one system and this co-operation should give better system performance. To make co-operation of these parts possible and effective – one should take integration and multiobjective optimization into account on each stage of development process. Otherwise, parts of different nature will not co-operate properly and effect of synergy can be lost. The mechatronic approach to design can be based on application of virtual prototyping of system using model simulation [3]. Models simulation can be used for prototyping function and architecture of a system.

In the context of HM-FDIR system design, mechatronic approach helps to achieve better integration of HM-FDIR system with structure under monitoring, mainly simulation techniques are used for functional and timing validation of designed system [4, 5]. The structures with HM-FDIR system have heterogeneous nature, i.e. they consist of analog/digital, hardware/software and/or electrical/mechanical components which is tackled by simulator coupling. Using heterogeneous simulation the components of HM-FDIR systems can be optimize to obtain better performance under different operating regimes [6, 7, 8, 9, 10, 11].

In vibration monitoring and control with application to HM-FDIR there are some problems which can be solved using mechatronic design approach. One of this problem is identification of operational forces which are reasons of high amplitude vibrations. The forces during operation are very difficult to measure directly [12, 13, 14, 15]. The smart sensor dedicated for load measurement is designed and implemented.

The sensors measure directly system responses and employing inverse identification algorithm allows to estimate forces which are reason of measured vibration (responses).

The second problem related to high amplitude vibration is repairing of system which is damaged because of unbalancing. For balancing, machines with high vibration level are stopped and balanced with iterative procedure. This is costly and time consuming procedure especially if turbines or generators require balancing. The automatic balancing system can be very helpful to solve this problem during operation.

Mechatronic approach has been applied by author for smart sensor and automatic balancing unit design and implementation. Methodology of design and obtained solutions are presented in this paper.

2 Mechatronic Design Idea

Mechatronic design idea is based on concept of modeling of design interdisciplinary system, integration of components with different physical nature and treating of all system components with the same weights during design process. Required in design process models can be obtained using bond graph based method [9], block diagram method [10] or UML based approach [11]. In presented project block diagram modeling with application of SIMULINK® software environment has been employed and developed.

In mechatronic design four main stages can be distinguished: conceptual design, virtual prototyping, details design and physical prototyping. The scheme of the proposed procedure is shown in Fig. 1.

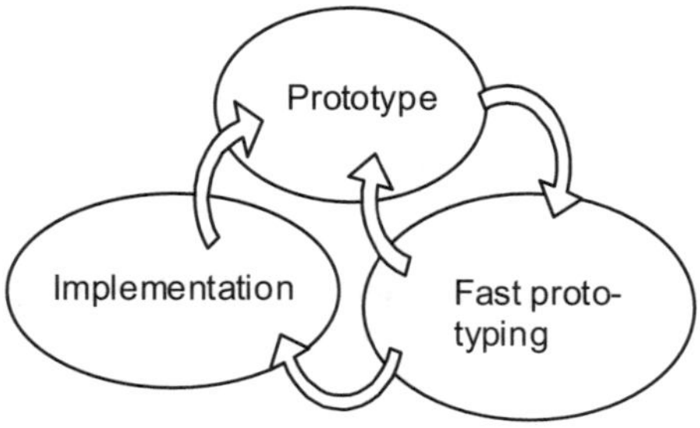

Fig. 1. Fast prototyping procedure

Virtual prototyping is based on a model simulation and can be use for testing of function and architecture. Physical prototyping of mechatronic systems is one of the most important stages in mechatronic design concept. At this stage the concept of designed structure and models of the system are verified employing experimental testing. The fast prototyping procedures consist of real-time simulation of mechanical, electronic and control parts of the system are employed to test components synergy in product design [20, 21]. The prototyping can be done; on architecture level, to test architecture of the system and functional performance or on system level, to test and develop interfaces between elements with different physical nature or on single component level, to optimize their structures. The prototyping procedures can be realize in one software environment using the same simulation model.

The scheme of virtual and physical prototyping software architecture applied in described projects is shown in Fig. 2.

Central role of this software architecture plays SIMULINK® simulation environment [22] in which the model of the system and its controller are modeled using block diagram methods. The mechanical part of designing system is modeled using CAD interface build in MSC VISUAL NASTRAN [23] software or in more complex cases in LMS Virtual Lab. Motion [24] software. Both multibody modeling software systems have possibility to transfer models, automatically to s-functions compatible with SIMULINK block diagram structure. Analysis and simulation task can be done using nonlinear simulation engine build in SIMULINK environment. Automatic c-code generation and real-time simulation help to test prototype system performance. For fast prototyping of control systems the dSPACE [25] hardware and software solutions with digital signal processors on board is proposed to use.

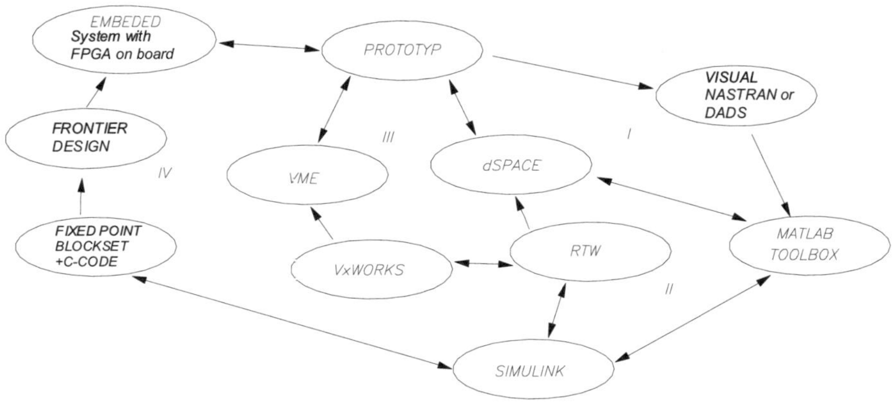

Fig. 2. Applied software for mechatronic system prototyping and implementation

Controller implementation phase of the mechatronic design process can be realize using three possible solutions; industrial computer, embedded controller or ASIC chip. The newest technology of ASIC chip solutions includes; field-programmable gate arrays (FPGA), field-programmable interconnect components (FPIC), and field-programmable analog arrays (FPAA) [6, 7, 8, 9] . In presented demonstration example the FPGA based smart sensor implementation is applied. FPGA are reconfigurable digital device, which mainly consist of an array of programmable functional blocks and I/O-pads as well as a network of reconfigurable interconnections. A new generation of FPGA chip offers dynamic reconfigurability that means possibility of the device to reconfigure during operation.

In the paper prototyping and implementation of smart sensor and automatic balancing system are presented. In the first case ASIC/FPGA solution is employed but industrial computer based controller has been implemented for second one.

The software environment shown in Fig. 2 consists of four design loops. The first loop is modeling and simulation loop for virtual prototyping on functional and architecture level. The second loop realizing fast prototyping procedure with automatic C-code generation which can be run on DSP platform for real time simulation of controller. In the third loop, based on generated and updated during fast prototyping C-code, control system is implemented using industrial computer hardware platform. But based on FPGA chip solution controller can be design using software shown in loop four. The procedure for FPGA application is detail described in the next section.

3 Operational Load Measurements - Smart Sensor Design

In order to identify operational load, based on process parameters or system response measurements, various methods can be used [12, 13, 14, 15, 16, 17, 18, 19, 20, 21]. This task is also known as an inverse identification problem [18, 19, 20, 21]. Methods that can be used to solve inverse identification problem are classified into three main categories:

- deterministic methods,
- stochastic methods,
- artificial intelligence based methods.

Two main techniques belong to the group of deterministic methods can be distinguished:

- frequency domain methods [13, 14, 17],
- time domain methods [12, 14, 19, 20, 21].

This classification is based on the signal processing methodology, which is applied to experimental data processing in order to perform estimation of loads parameters.

Basic methods of excitation identification have been formulated for linear systems for which assumptions about small damping and parameters stationarity are valid. Frequency domain methods require information about FRF (Frequency Response Functions) of investigated structures and spectrum of system responses measured during operation. Based on this information, spectrum of exciting forces can be estimated.

Similar methods are formulated in the time domain, using relation between excitation and system responses in a form of convolution. An iterative formula for calculations of excitation forces that act on mechanical structures is proved based on properties of the Toeplitz matrix [13].

Identification methods shortly described above can be used for linear systems only. For the analysis of both linear and nonlinear systems, methods based on minimization of given objective function can be employed. Usually, a least square error between simulated and measured system response is used as objective function in this identification methods. The dynamic programming optimization method formulated by Bellman [13, 19, 20] is commonly used for minimization of objective function in order to estimate excitation forces. Some examples of the application of these methods are shown in [13, 19, 20]. Similar approach based on genetic algorithms is presented by author in [21].

Statistical approach is formulated with the use of statistical models of relation between system response (or process parameters) and operational loads. The approach based on the regression model of relation between loads and flight data (data recorded using standard flight recorder) is presented for a helicopter structure in [16, 17].

3.1 Smart Sensor Design Concept

The smart sensor for diagnostic purpose should have extended possibility of signal and data processing, specialized for given physical parameter estimation. This creates specific requirements for such design. Transducer should have processor on board. The processor can be applied for data processing and be programmed or learned. Transducers that have built a preprocessing electronics (signal conditioning), processor and communication capability are called smart sensors [26, 27].

Proposed solution has possibility to be programmed externally from operator workstation that function of sensor can be changed by user, if internal software will be changed. The software of smart sensor commonly is implementation of physical phenomena model or its inversion. In a case of intelligent algorithm, mainly neural networks are implemented inside transducers [27]. The design procedure [28, 29, 30] specific for smart sensor solutions schematically is shown in Fig. 3.

First step in proposed procedure is transformation of employed model from continuous time simulation to fixed-point. Discrete time form is accomplished in Simulink and allows for testing signal dynamics and behavior of fixed-point

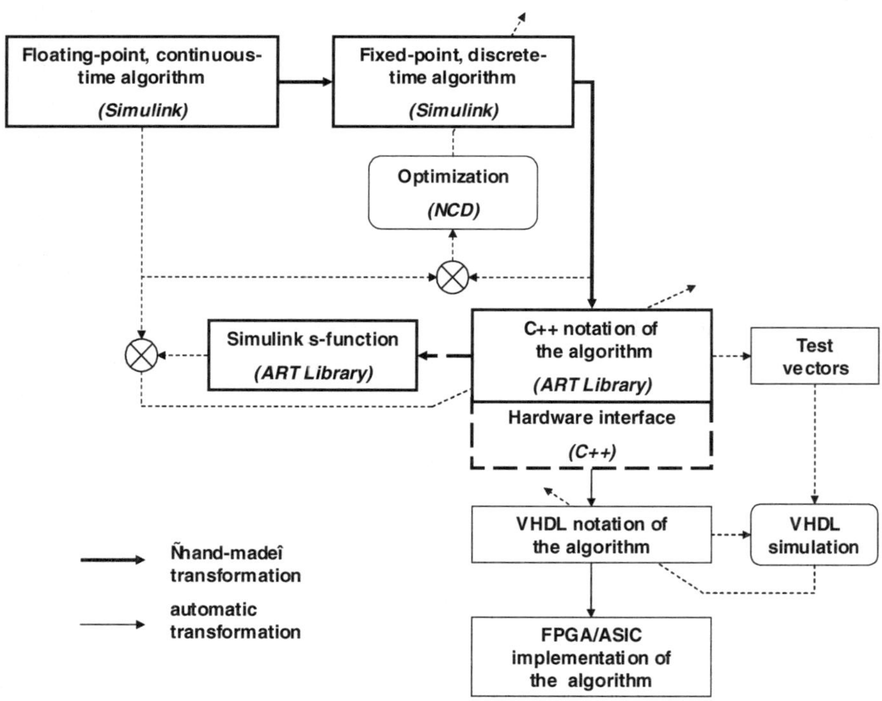

Fig. 3. Procedure of neural network based algorithm implementation in ASIC/FPGA

computation. The next stage uses software from Frontier Design [28, 29]. Its A|RT Library is a library for several C++ implementations, providing fixed-point classes, which allows for simulation of fixed-point calculations in C++. The C++ code utilizing fixed-point classes can be then synthesized into VHDL or Verilog [30] using A|RT Builder. Useful C++ code of an algorithm can be encapsulated by SIMULINK interface code to create so-called s-function. Such an s-function defines behavior of a new block, which can be included into each system model formulated in SIMULINK [31].

Concurrently, the same s-function source can be synthesized by A|RT Builder into VHDL. Generated VHDL code depends only on C++ programming style. By choosing a suitable style, resulting chip topology can be optimized for speed or for capacity. It provides a high degree of control over generated VHDL code, at the same time freeing from hand coding. Generated VHDL code of an algorithm, together with VHDL description of necessary peripheral interfaces is then compiled into gate level netlist, which describes chip architecture using gates, flip-flops and other primitives available in a particular technology [33]. Later, a gate level netlist need to be fitted, which means, that physical chip resources are to be allocated to each primitive in the netlist. In this case, Synopsys FPGA Express [34] was used as a compiler and Altera Quartus II was used as a fitter [35], which ends the implementation procedure.

Smart sensors algorithms can be develop in proposed procedure using modeling and simulation tools from wide range of SIMULINK toolboxes software family, e.g. NN Toolbox, NCD Toolbox, Fixed Point Blockset, etc. [36, 37, 38].

Useful C++ code of an algorithm can be encapsulated by SIMULINK interface code to create so-called s-function. Such an s-function defines behaviour of a new block, which can be included into each system model in SIMULINK. Concurrently, the same s-function source can be synthesized by A|RT Builder into VHDL [38, 39].

Important advantage of A|RT Builder is its transparency. It means that generated VHDL code depends only on C++ programming style. By choosing a suitable style, resulting chip topology can be optimised for speed or for area, thanks to synthesizing such more sophisticated data flow control techniques as pipelining or resource sharing. It is well known what VHDL code is generated for each C++ statement or expression and all identifiers are preserved. It provides a high degree of control over generated VHDL code, at the same time freeing from hand coding. Moreover, it makes possible to link VHDL control algorithm notation with other VHDL code, e.g. describing hardware interfaces with peripheral devices as Analog-to-Digital Converters (ADC) or Digital-to-Analog Converters (DAC) as well as calculation time frame (-s).

Generated VHDL code of an algorithm, together with VHDL description of necessary peripheral interfaces is then compiled into gate level netlist, which describes chip architecture using gates, flip-flops and other primitives available in a particular technology. Later, a gate level netlist need to be fitted, which means, that physical chip resources are to be allocated to each primitive in the netlist. In this case, Synopsys FPGA Express [34] was used as a compiler and Altera Quartus II was used as a fitter.

In a case of necessity of manual modification of VHDL code, some way of checking its consistency is needed. Testing can be performed with the use of VHDL simulators. Creation and verification of test vectors needed for VHDL simulation is supported by A|RT Library.

3.2 Concept of Smart Sensor for Operational Force Measurements

In order to measure load of particular structural components of mechanical structure, special design of sensor is needed. The sensor should extend possibility of signal and data processing, specialized for load vector estimation. This creates specification requirements for such a design. Transducer should have a microprocessor on board. The processor is applied to data processing and is programmed or learned to communicate with data acquisition system. The main functions of such sensors are as follows:

- collect response signals from local transducers,
- perform data processing,
- estimate load vector parameters,
- send information abut the load to the acquisition system.

The scheme of such a sensor design is shown in Fig. 4.

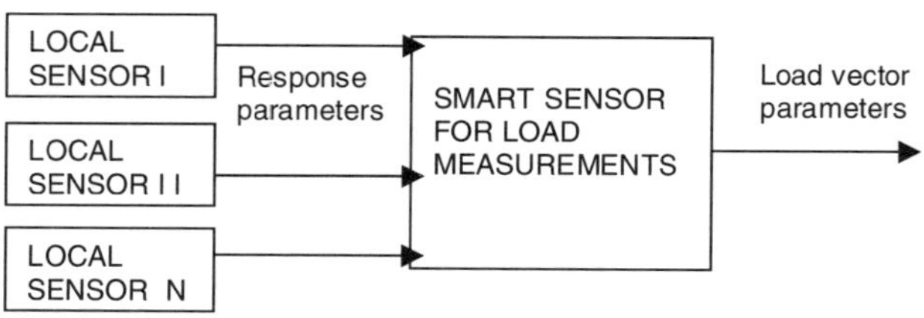

Fig. 4. Scheme of the smart sensor concept for load measurements

In a case of acceleration measurements at several points on a structure measured as response parameters, typical accelerometers can be used as local transducers. Analog signals from local sensors are collected by signal acquisition module build in smart sensor structure. Signals are converted to digital form and recorded in transducer memory buffer, the next step of the procedure implemented in a smart sensor is data processing with the use of system model in a form of a deterministic formula or an intelligent algorithm, which represents mapping of response signals to load vector. In a case of an intelligent algorithm, mainly neural networks are implemented inside transducer. This neural network needs to be trained before using it to load vector identification. This requires one or more additional inputs to measure load vector during learning phase. The learning process can be performed outside the sensor. Ready-to-use network is loaded to the sensor memory. The solution based on microprocessor or ASIC solution can be implemented. The scheme of the proposed sensor internal structure design is shown in Fig. 5.

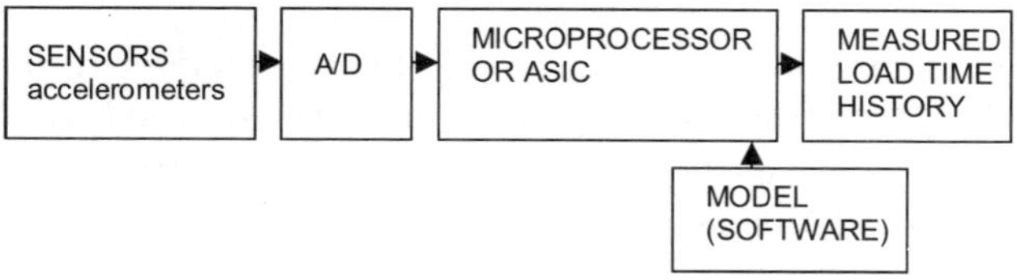

Fig. 5. Proposed smart sensor design

Smart sensor for load measurements based on neural networks algorithm and ASIC/FPGA hardware platform is designed, implemented and tested in the Department of Robotics and Machine Dynamics at the University of Mining and Metallurgy.

3.3 A Hardware of the Smart Sensor

From the hardware point of view, the intelligent sensor consists of several parts, shown schematically in Fig. 6. The actual sensing elements – primary transducers – convert mechanical quantity into electrical one. The resulting signals usually require some kind of amplification or buffering and antialising filtration. Then they need to be transformed to a sequence of digital samples by means of analog-to-digital converters (ADCs). Whole digital signal processing takes place in a single integrated circuit, namely ASIC, or especially during prototyping, in its user programmable counterpart (FPGA). The output of intelligent sensor can be analog, which requires a digital-to-analog converter, or digital. In the latter case, some mechanisms of data exchange – communication interface – have to be implemented in the same ASIC/FPGA chip [30].

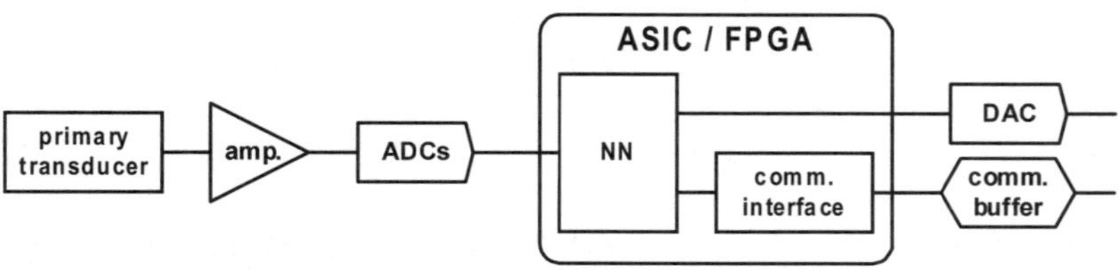

Fig. 6. Intelligent sensor hardware architecture

Nowadays, piezoelectric accelerometers are used as primary transducers, Altera APEX 20KE family FPGA chip [40] as the primary unit, and the system output is of the analog type.

3.4 Testing Performance of Smart Sensor

The performance of the smart sensor has been evaluated by carrying out an experiment. The output of the sensor has been recorded together with a real measured force in the experimental rig shown diagrammatically in Fig. 7. Three fragments of testing signals are shown in Fig. 8.

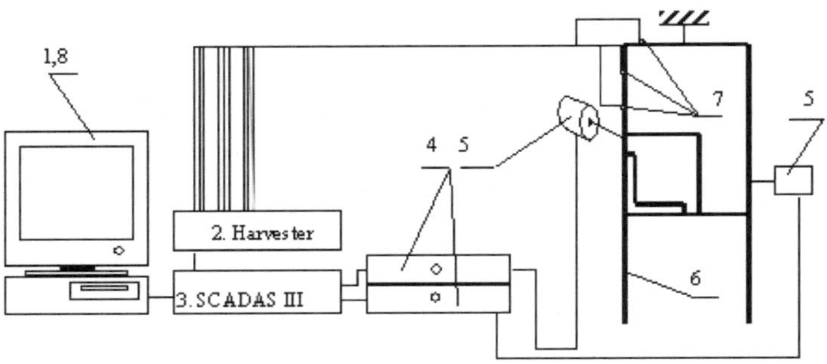

Fig. 7. Experimental rig for verification of developed algorithms

The whole application, neural network calculations and peripheral interfaces, occupies about 80% of logic elements and about 8% of the memory in EP20K100E FPGA chip, leaving enough resources for future realization of a communication protocol and interfaces. Working at 6.25MHz, system carries out neural network calculations in 45μs. After addition of 5μs for servicing external devices, it gives maximum sampling rate of 20kHz. Faster clock (20MHz is permissible at the moment) would allow to achieve 65kHz sampling frequency,

A good agreement between directly measured loading force and estimated by the smart sensor one can be observed. One more advantage of application of the smart sensor presented in the paper is possibility of determining loads in real time, which is very important in case of applications in monitoring and control systems.

4 Mechatronic Device for Automatic Shaft Balancing

Vibration related problems due to unbalance in rotating equipment can be solve using automatic balancing device installed on rotating shaft. Active automatic balancing uses a device which is permanently installed on a piece

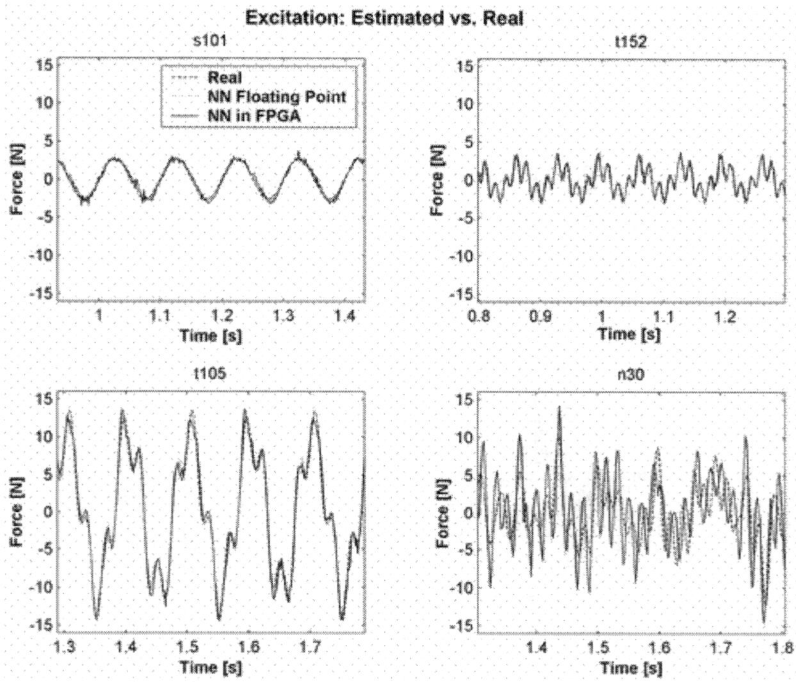

Fig. 8. Verification test results

of rotating equipment. The balancing system in continuous way monitor vibration levels of bearings and makes balance mass corrections automatically during operation. Until now, in case of too high vibration level, most balancing methods required to stop the machinery and fix an additional correction mass in special balancing planes. In vast majority of cases, balancing procedures must be carried out several times in order to achieved satisfactory level of vibration.

On-line rotor balancing requires a special controlled devices or mechanisms which can change angular and radial position of correction mass with respect to the rotor shaft rotation. This kind of correction mechanisms should have two degrees of freedom. Dynamic balancing of rigid rotors requires two correction planes, which means that a complete correction device has four degrees of freedom. There is several different solutions of the active balancing mechanism [41, 42, 43].

But new innovative mechatronic active balancing device shown schematically in Fig. 9 has been designed [41].

Since mechanical and control systems of balancing device are complicated, the balancing device should be used only in rotors where unbalance is high and changes relatively often during operation.

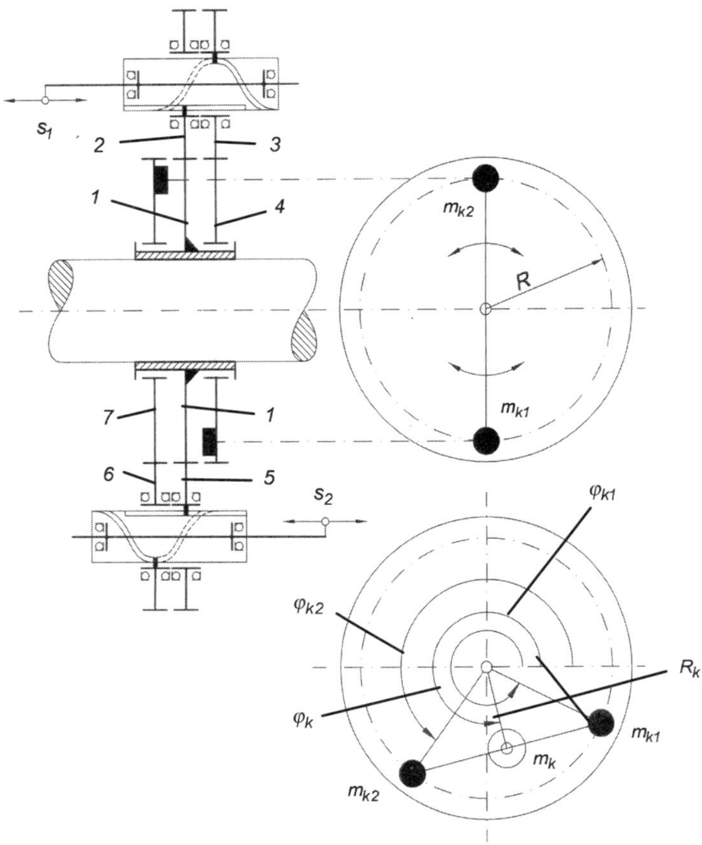

Fig. 9. Scheme of designed active balancing mechanism

Controller is an integral part of the active balancing device of the rotor systems. The main tasks of the controller are: rotor vibration measurement, calculation of the correction masses placement and control of the correction masses movement. The scheme of controller is shown in Fig. 10. The main element of a kinematic chain is toothed gear. The proper operation of the toothed gear can cause some problems because it requires cooling and lubrication. Toothed wheels mounted on rotors shaft have the same speed as rotor, and rotational speed of coupling wheels is dependent on the transmission ratio.

Considering the technical problems mentioned above, the toothed wheel has been replaced by toothed belt gear. Toothed belts provide high geometrical accuracy and they are less vulnerable to assembly errors.

Presented here technical concept of a correction device meets technical requirements for rotor shaft active balancing devices, and has been used to design prototype. Proposed active balancing method permits for on-line balancing.

Methods of calculating correction mass's position used in active balancing procedure are similar to ones used in conventional methods and can be divided into two categories:

- modal methods,
- influence coefficient method.

Modal methods, based on a rotating structure models, determine experimentally the disturbing unbalance associated with a specific mode.

The influence coefficient method uses an experimental model that represents the machine sensitivity to unbalances. Determining the influence of the unit correction mass on the machine vibrations is the basis of this method. In order to estimate this influence, a test mass is placed in each balancing plane. In the following step, the system response on a changed mass distribution is measured and, on the basis of these measurements, the influence coefficients are estimated.

In a proposed solution the influence coefficient method was used. This is an empirical methods and it doesn't require a-priori knowledge on the rotating system.

In original formulation of influence coefficient method, the coefficients of the j-th plane H_{ij} are calculated based on experiment with additional mass W_{cij} added in the balancing plane i-th and measured vibration level:

$$H_{ij} = \frac{V_{cwi} - V_i}{W_{cwj}} \tag{1}$$

where: V_{cwi} – the vibrations measured in i-th measurement plane after addition of test weight W_{cwj}, V_i – the vibrations measured at this plane before addition of test weight.

After measuring vibration and calculating of influence coefficients in j-th plane trial mass is removed and placed in next plane. If the matrix H_{ij} is fully determined, the unbalance correction vector W_c can be determined from the formula:

$$W_c = -H^{-1}V \tag{2}$$

Formula (2) can be applied directly if matrix H is square matrix, e.g. number of measuring planes is equal to correction planes, and otherwise, pseudo-inversion should be applied.

During active balancing realize on designed mechanism no mass is added or subtracted from the system, correction force is generated as results of changing in position of satellites at each correction planes. For unbalance correction with W_c and angle Θ, angle of satellites position change are obtained from formula:

$$\alpha = \arccos \frac{W_c}{2w_j} \tag{3}$$

where w_j is unbalance generated with j-th correction mass.

Formula (3) allows controlling correction satellites position during shaft operation.

4.1 Synthesis of the Satellites Control

Controller is an integral part of the active balancing device of the rotor systems. In this chapter software and hardware of the controller will be described.

The main tasks of the controller are:

- rotor vibration measurement,
- calculation of the correction masses new positions,
- control of the correction masses movement to new positions.

It is very difficult to measure a shaft vibration in a bearing directly. Usually vibration measurements are taken by sensors mounted on the bearing housing. Calculation of the correction mass position requires vibration component synchronous to the rotation (1x) of the shaft, it requires an additional sensor, keyphasor. This sensor is also required to correlate vibration level with a correction mass position.

Influence coefficients and correction mass positions are calculated by the controller with the use of the algorithm described in the previous section.

Positions of the correction masses are changed by movement of the control shafts. The main task of the controller is changing correction masses position to the desired one, by controlling shaft drives. During correction mass positioning, the increase in the unbalance (vibration level) is avoided using alarm signal, which stopped rotation of control mechanism. Since the range of changes in correction mass position is limited due to the mechanism design, in some cases it is impossible to obtain the desired position in a safety way. Such cases should be detected and indicated. Knowledge of the current position of the correction mass is critical for safe work of the correction device. Because calibration can't be taken after each start-up of the device, absolute position sensors is used.

Vibration level measurements, correction masses positions, influence coefficients and unbalances history are available for the user.

Controller for mechatronic balancing unit consist of two main parts; hardware and software part.

Hardware Part of Controller

Scheme of controller architecture is shown in Fig. 10. The controller contains the following parts:

- vibration pick-up,
- keyphasor,
- shafts position sensors,
- shafts drivers,
- other superior subsystems.

Vibrations are measured by accelerometrs mounted on rotor's bearing housing, signals is taken from diagnostic and safety systems. In some cases additional sensors are needed (if more balancing planes are planed). Two kinds of sensors can be used for measurements: vortex sensors and accelerometers. Both types have standard voltage outputs, differing only in its range. For a single plane measurements two vortex sensors or one accelerometer are required.

Position of correction masses is measured, in proposed solution, with absolute position sensors equipped with serial digital outputs.

In order to move a coupling shaft electric motors with screw gears is used, in presented case brushless asynchronous motors. These types of motors are provided with amplifiers with internal feedback as servomotors.

Standard asynchronous serial connection is recommended to lead out vibrations and correction masses position signals.

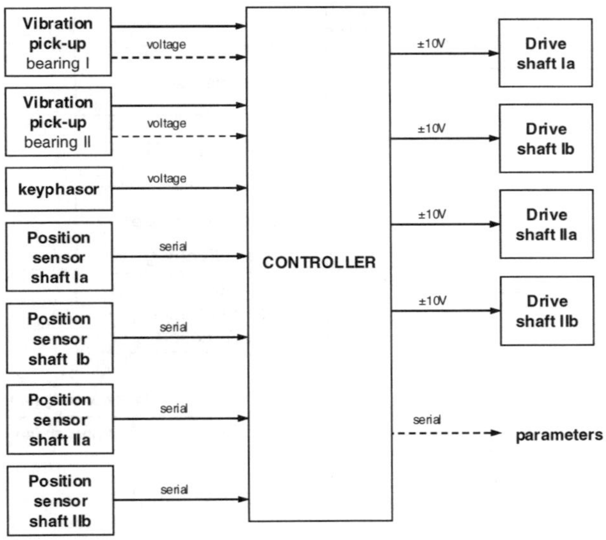

Fig. 10. Controller with external devices

The controller is equipped with the following input-output systems:

- two (four, when vortex sensor are used) voltage inputs with adjustable ranges and sampling frequencies, with minimum resolution 12 bits and minimum sampling frequencies 500 times higher than rotor rotation frequency,
- single counter-clock input with resolution 500 times higher than rotor rotation period, with adjustable range of input voltage, for keyphasor,
- four serial digital inputs for shaft position sensors,
- four voltage outputs ± 10V with 10 bits resolution, for shaft drivers,
- serial communication interface for data transmission.

Modular industrial computer with input/output boards should be used as controller hardware. Controller can be easy adapted for using different sensors by changing type of input/output boards. Because it is assumed that system is quasi-static, real-time calculations are not required. Hardware allows for vibration data acquiring for period which allows for proper data processing and theirs synchronization with keyphasor signal.

Software Part of Controller

Data processing is carried out according to the scheme shown in Fig. 11.

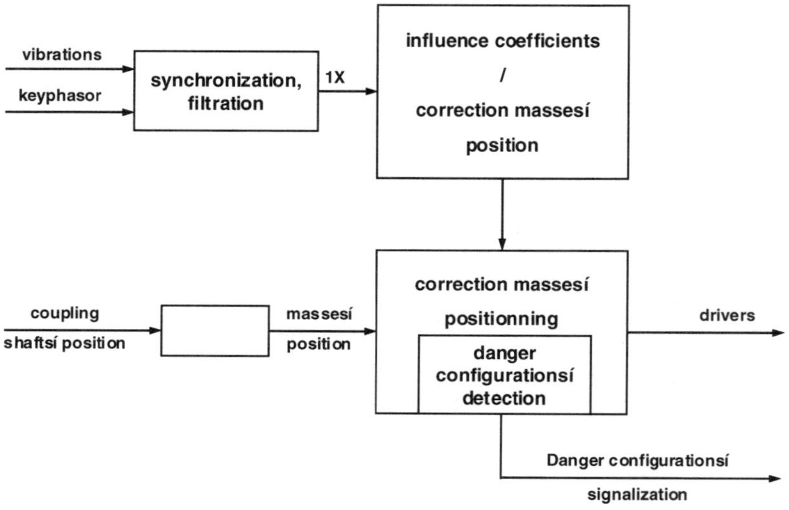

Fig. 11. Scheme of data processing

After synchronization with keyphasor signals, vibration data are filtrated to separate amplitude and phase of 1x component, which is used for influence coefficients and correction masses position calculations. Using desired and currently correction masses positions, measured by shaft position sensors, drivers signals are generated. System checks if the desired position of correction masses can be reached from a current shafts position safety, if not, the shafts stops and system generates an additional alarm signal. Vibration level is controlled during correction masses movement from their initial position towards their desired, correction position and if vibration level increases over acceptable level, masses movements are stopped or even they are placed in their initial positions.

Acquisition of vibration data ensure their logging and synchronization with keyphasor signals. Proper filtration process requires information about a rotor rotational speed. These tasks have the highest priority. Data filtration and correction masses position calculations are carried out off-line and relatively seldom, with lowest priority. Masses positioning should be carried out in real time but their movements are slow so it isn't a time critical task.

Balance correction device after installation and after each re-assembling requires calibration of keyphasor with correction masses position and vibration sensors.

Influence coefficients, the last stable positions of correction masses with their vibration levels are kept in controller's permanent storage in case of the controller shut down. The controller has been designed employing mechatronic approach described in section 2 [44, 45].

4.2 Active Balancing System Controller Design Procedure

The first step of controller design was simulation of its model which has been prepared using SIMULINK software environment. The SIMULINK scheme of proposed controller is shown in Fig. 12.

The simulation results help to assess performance of the control scheme and update parameters of controller. Final simulation results are shown in Fig. 13.

Parameters which are changing during simulations are feedback loop gains and weighting mass position parameters. The controller changing angular position of balancing mass based on vibration amplitude and phase measurements.

In Fig. 13 the results for one and two planes unbalance corrections using designed controller are shown as time plots of vibration measured by sensor located on bearings cover. Associated with this vibration time plots positions of correction mass are shown in Fig. 14.

The scheme in Fig. 12a presents simulation model of controller with rotor system and and sensors, but scheme in Fig. 12b shows controller internal structure which is developed and updated during simulation process. Bearing covers vibrations for considered rotor system are very sensitive on masses position changing. As it can be noticed from obtained results; designed automatic balancing system works very effectively, even small corrections of control shaft positions the balancing can be changed sufficiently.

4.3 Conclusions

As it is shown in the paper mechatronic approach gives many advantages in design of HM-FDI systems and helps to achieve better integration between monitored system and monitoring and diagnostic system.

Presented mechatronic design approach presented in the paper, based on virtual and fast prototyping procedure shorter design time and automates design and programming tasks.

Designed smart sensor for health monitoring applications can be very easily (in automate way) modify to measure different physical value and can be

a)

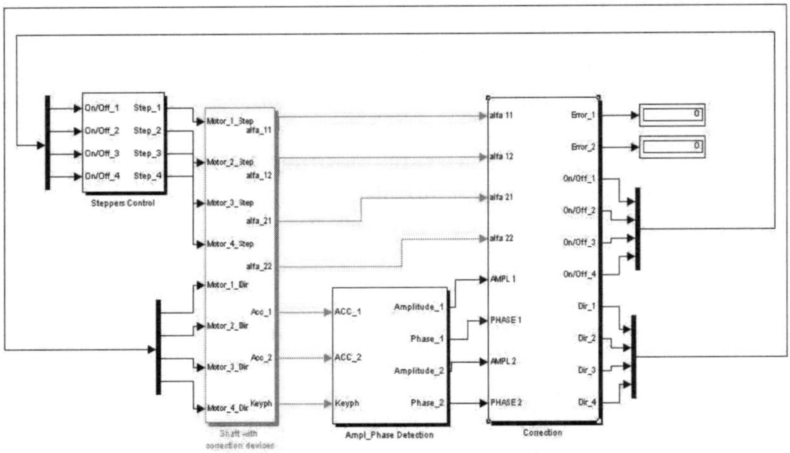

b)

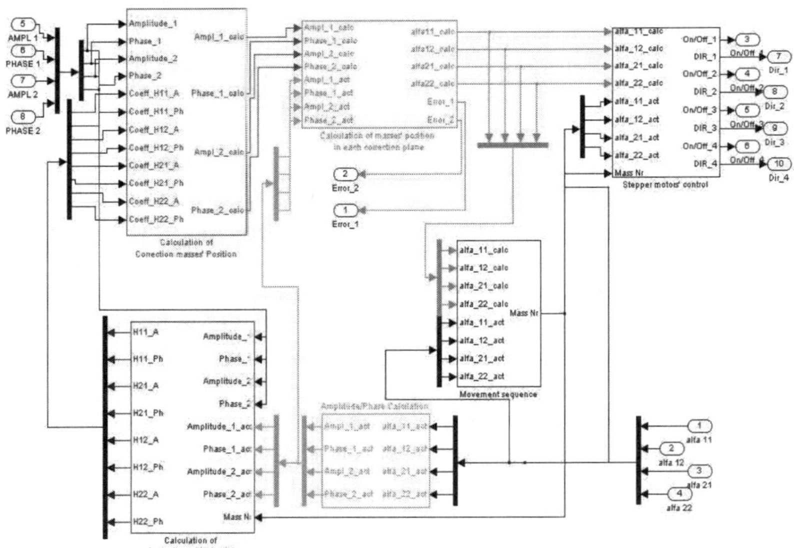

Fig. 12. SIMULINK scheme of designed controller

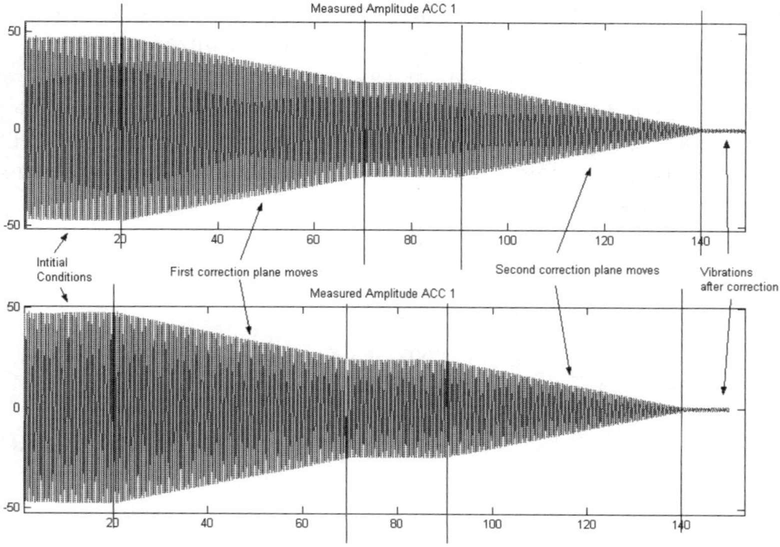

Fig. 13. Simulation results for different control task

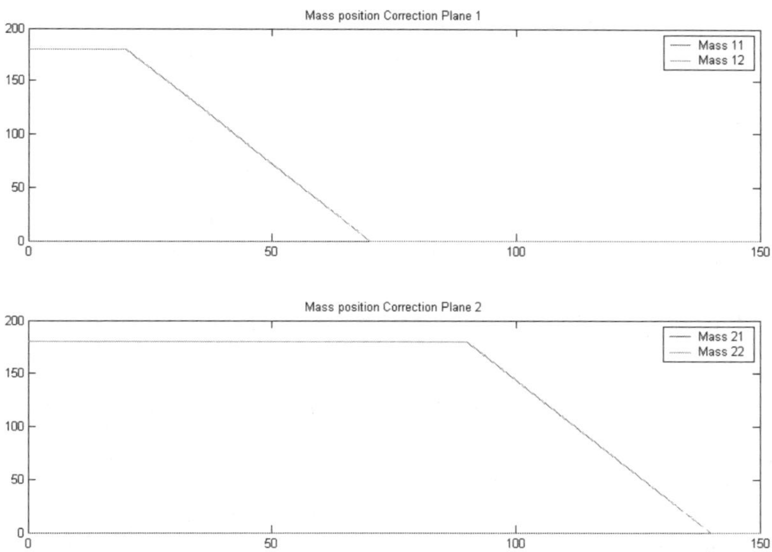

Fig. 14. Time history of correction mass positions changes due to control signals

applied for different HM-FDI system. The flexibility is achieved because of using FPGA programmable electronic chip with specialized software.

The example of automatic balancing system shows possibility to build HM-FDI system, which can correct occurred and detected damage automatically. Designed balancing system can be applied for steady state and transient operation of rotating machinery. The system is now modifying to be applied for steam and gas turbines.

References

1. Caccavale F, Villani L (Eds.) (2003) Fault diagnosis and fault tolerance for mechatronic systems. Springer, Berlin
2. Isermann R (2003) Mechatronic Systems. Springer, London
3. Uhl T (Ed.) (1999) Selected problems in mechatronic design. Robotics and Machine Dynamics Department Press, University of Science and Technology AGH, Krakow
4. Uhl T, Barszcz T, Hanc A (2003) Mechatronics in design of Monitoring and Diagnostic Systems. Key Engineering Materials 245-246:381-390
5. Petko M (2003) Fast prototyping in development of diagnostics and monitoring systems. Exploitation Problems of Machines 38(3):95-114
6. Harpel H-J, Glesner M (1998) Rapid prototyping of real-time information processing units for mechatronic systems. Real-Time Systems 14:269-291
7. Rowson A (1994) Hardware/software co-simulation, Proc. Of the 31st Design Automation Conference, San Diego, USA, pp 129-134.
8. Glesner M, Kirshbaum A, Renner FM, Voss B (2002) State of the art in rapid prototyping for mechatronic systems. Mechatronics 12:987-998
9. Damic V, Montgomery J (2003) Mechatronics by Bond Graphs. Springer, Berlin
10. Uhl T, Rod J (1999) Modeling and simulation of Mechatronic systems, In: Selected problems in mechatronic design, T. Uhl (ed.), Robotics and Machine Dynamics Department Press, University of Science and Technology AGH (in Polish)
11. Mrozek Z (2002) Computer assisted design of mechatronic systems. Sci. Bul. Cracow Univ. Tech., 1
12. Uhl T (1999) Trends and progress in monitoring and diagnostic systems. PAK 4, (in Polish)
13. Giergiel J, Uhl T (1989) Identification of the input impact type forces in mechanical systems. The Archives of Transport 1(1)
14. Uhl T (2001) Identification of loading forces in mechanical systems using genetic algorithms, Proc. of AIMECH 01, Gliwice
15. Haas DJ, Milano J, Flitter L (1995) Prediction of Helicopter Component Loads Using Neural Networks. J. American Helicopter Society vol. 1
16. Zion L (1994) Predicting fatigue loads using regression diagnostics, Proc. American Helicopter Society Annual Formu, Washington D.C.
17. Lisowski W, Mendrok K, Uhl T (2001) Identification of loads basing on output signal measurement, Proceedings of V Conf. on Experimental Methods in Machine Design, Wrocaw- Szklarska Poręba, (in Polish)
18. Busby HR, Trujillo DM, Solution of an invesrse dynamics problem using an eigenvalue reduction technique. Computer & Structures 25(1)

19. Simonian SS (1981) Inverse problems in structural dynamics. Int. J. Num. Meth. Engng. 17:357-365
20. Trujillo DM (1987) Application of Dynamic programming to the general inverse problem. Int. J. Num. Meth. Engng. 23:613-624
21. Uhl T, Pieczara J (2003) Identification of operational loading forces for mechanical structures. Archives of Transport 16(2)
22. SIMULINK User's Guide (1999) The MathWorks Inc., Natic.
23. `www.msc.com`
24. `www.lms.be`
25. `www.dspace.de`
26. Frank R (2000) Understanding Smart Sensors. Artech House, Norwood
27. Uhl T, Petko M (2002) Smart Sensor for Operational Load Measurements. J. Theor. Appl. Mech. 40(3):797-815
28. Petko M (1999) Mechatronics product realization (in Polish), In: Mechatronics design (T. Uhl (Ed.)), KRiDM AGH, Kraków
29. Petko M (2001) Implementation of control algorithms in ASIC/FPGA. Mecatronics. Proceedings 5th Franco-Japanese Congress & 3rd European-Asian Congress of Mechatronics pp. 70-74, Besancon
30. Petko M, Uhl T (2001), Embedded controller desing-mechatronic approach Proc. Second International Workshop on Robot Motion and Control RoMoCo'01, Politechnika Poznańska, Bukowy Dworek, pp.195-200
31. Uhl T, Mrozek Z, Petko M (2000) Rapid control prototyping for flexible arm. 1-st IFAC Conference on Mechatronic Systems, Preprint, 2:489-494, Darmstadt
32. Jang J-SR, Sun C-T, Mizutani E (1997) Neuro-Fuzzy and Soft Computing: A computational approach to learning and machine intelligence. Prentice Hall, Upper Saddle River
33. Perry DL (1998) VHDL, McGraw-Hill
34. FPGA Compiler II / FPGA Express VHDL Reference Manual (1999) Synopsys, Inc.
35. Neural Network Toolbox User's Guide (1999) The MathWorks Inc., Natic
36. Fixed-Point Blockset User's Guide (1999) The MathWorks Inc., Natic
37. Nonlinear Control Design Blockset User's Guide (1997) The MathWorks Inc., Natic
38. A|RT Library User's & Reference Manual (2000) Frontier Design Inc., Danville
39. A|RT Builder User's and Reference Documentation (2000) Frontier Design Inc., Danville
40. APEX 20K Programmable Logic Device Family (1999) Altera Corp.
41. Manka M, Felis J, Petko M, Uhl T (2003) The new method of automatic balancing during operation. Bull. Univ. Sci. Tech. AGH, 22(3):347- 354
42. Furman BJ (1982) A new theramly controlled, non contact rotor balancing method. J. Mechanical Design 116:823-832
43. Van den Vegte J (1981) Balancing of flexible rotors during operation. J. Mech. Engng. Sci., vol. 23(3):257-261
44. Hanselmaan H (1993) Hardware in the loop simulation as a standard approach for development, customization and production test. SEA Paper no. 930207
45. Uhl T (1996) Fast prototyping - a new tool for mechatronic system development. PAK, 4:34-41

Printing and Binding: Strauss GmbH, Mörlenbach